CAMBRIDGE COUNTY GEOGRAPHIES

General Editor: F. H. H. GUILLEMARD, M.A., M.D.

DORSET

Cambridge County Geographies

DORSET

by

ARTHUR L. SALMON

With Maps, Diagrams and Illustrations

Cambridge:

at the University Press

1910

CAMBRIDGE UNIVERSITY PRESS
Cambridge, New York, Melbourne, Madrid, Cape Town,
Singapore, São Paulo, Delhi, Mexico City

Cambridge University Press
The Edinburgh Building, Cambridge CB2 8RU, UK

Published in the United States of America by Cambridge University Press, New York

www.cambridge.org
Information on this title: www.cambridge.org/9781107614024

First published 1910
First paperback edition 2013

A catalogue record for this publication is available from the British Library

ISBN 978-1-107-61402-4 Paperback

CONTENTS

ILLUSTRATIONS

ILLUSTRATIONS

MAPS

1. County and Shire. The Origin of Dorset.

The earliest divisions of England were tribal. The tribe was a large family, or cluster of families more or less connected, under one head or chieftain; and it occupied as much territory as it was able to hold by the strong arm. The delimitation of a tribe's frontiers was usually assisted by nature; hill-ranges, forests, rivers, or great swamps forming the natural boundaries that divided one tribe from another. Of course there had been a time before even the tribe came into existence, when men counted by units instead of in aggregates; but this was before the dawn of civilisation and political government. When the Romans came they did not interfere with these tribal divisions, but they instituted an imperial supremacy very similar to that which we at present hold in India. The Saxon invasion had a different effect. Tribes of Saxons and Angles gradually spread westward and northward, never really reaching Cornwall or Wales, and to a considerable extent they adhered to the old tribal divisions, thus proving the force of the natural boundaries that in most cases had governed the

delimitation. But in time the kingdoms that they established overleapt the former frontiers, and ultimately
were parcelled into shires or counties. A few existing
counties still retain the old tribal contour and name;
Dorset is one of these. There are some names of counties
that explain themselves; thus, Sussex was the kingdom of
the South Saxons, Middlesex of the Middle Saxons, Essex
of the East Saxons; Suffolk, somewhat more vaguely,
was the land of the South People, and Norfolk of the
North People. But with Dorsetshire (the affix *shire*, as
with Somerset, is now generally dropped, and was no
part of the original name) it is different; its name is not
so obvious in meaning, and calls for historical explanation.

The name and formation of Dorset are more ancient
than the arbitrary division into shires. It may be well to
notice exactly what the word *shire* indicates. It was
anciently written "scir," and is derived from the Saxon
word *sciran*, to divide (*sciran* itself being a secondary
form of *sceran*, to cut off). A "scir" meant literally a
share, a shearing off. We have the same root in our
word sheriff, i.e. the shire-reeve—the representative of the
king for purposes of taxing and jurisdiction. In this sense
Dorset became a shire of the powerful kingdom of Wessex,
which at its strongest extended from Kent to Devon.
The term *county* was introduced by the Normans, who
used the title Count where the Saxons spoke of Ealdorman
or Earl; they borrowed it from the Romans, who gave
this designation (Latin *comes*) to governors or custodians
of special districts (such as the Count of the Saxon
Shore). The division into shires is usually attributed

to Alfred—though there is some doubt, as the term
"scir" certainly existed earlier. In Alfred's time we
find, for legislative purposes, the township (in more
important cases the *burh*) ; the Hundred, which is sometimes explained as denoting a hundred families, but more
probably signified a division that was expected to supply
a hundred men for military service ; and the shire or

Dorchester

county. The shire had its own "folk-moot" or court,
and the Hundred, if not originally, came later to have a
similar judicial assemblage.

Dorset then was a shire of the great kingdom of
Wessex, but for the origin of its name we have to go
back farther. Before the Saxons came the land had been
"occupied" by the tribe of Durotriges, much as we

"occupy" India. These people, a tribe of Celts from Gaul, had invaded and conquered a country that was still largely inhabited by Neolithic people—people of the New Stone Age, who had learned the use of metals. They were people of dark hair and long skulls, and have doubtless many living descendants in the west and southwest of England. It has been disputed whether the Durotriges belonged to the larger tribe of Dumnonii, who were Goidels (the Goidels or Gaels were the earliest Celtic invaders of Britain), or to the Belgae who were Brythons (the Brythons or Britons were the second Celtic invaders). The name Durotriges, which is a Latinised form, is of uncertain origin and meaning, but it appears to be connected with the Celtic *dwr*, or water, and *tre*, a dwelling; in which case it denotes the "dwellers by the water." Professor Rhys has suggested however that the "duro" really signifies door or gate. When the Saxons came they adopted the first syllable, and named their settlements here the *Dor-saetas*, and from this we have the present name of the county. The names of Wiltshire (Wil-saetas) and Somerset (Somer-saetas) were formed in similar manner. It is probable that the original *dwr* or water was the river Frome, flowing through Dorchester ; so that perhaps Dorset, like many other English counties, derives its name from its capital town.

2. General Characteristics. Position and Natural Conditions.

With the exception of Devon and Cornwall, Dorset is the most southern county in England, and Portland Bill is a kind of southward Land's End. It is a maritime county rather in regard to its extensive coast-line (75 miles) than to the maritime character of its industries. These

Poole Harbour

are not important. Shipping is on a scale that cannot compare either with Southampton and the Bristol Channel ports, or with the activities of the Mersey and the Tyne. Fisheries are far from prosperous, and only afford employment to a very few hundreds of the population. With the exception of Portland and Poole the harbourage of the coast is poor, and Poole harbour cannot receive vessels of large size. But there is ample recompense for these shortcomings in the agricultural and geological wealth of the shire. Its great fertility renders it one of our chief districts of successful farming. Lying somewhat away from the main national highways, protected in the past by its forests and hills, Dorset has suffered comparatively little for many centuries from internal strife or outward attack, and has been able to devote itself whole-heartedly to its grazing and grain-growing. Fine qualities of soil are assisted by an excellent climate, a sufficient but not excessive rainfall, and a prevalence of sunshine hardly excelled by any other county. With the exception of the coastal district south of Dorchester, the county lies within the English region of highest temperature. Though there are only two important rivers, the Stour and the Frome, the land is well watered by a number of smaller chalk streams from the Downs. More than three-quarters of the county's acreage lie under cultivation, the larger portion being in pasture, and even the more barren of the Downs afford excellent grazing. The short sweet turfage of the hills renders " Portland mutton " almost as famous as Welsh, while the luxuriance of the vales lends itself to the breeding of fine cattle. The shire's reputation for butter has become proverbial, " Dorset butter " being a household word ; but we must now regret that these capital home-products are under-sold by cheaper imports from the Continent. Dairy-life is a prominent feature in this corner of old Wessex, and most people know Dorset's speciality of " Blue vinny "

(or blue-veined) cheese.   Dorchester, the county-town,
is the chief agricultural centre and market-place, and its
activity in this department is thoroughly satisfactory.
The county can play its part well in producing the food
of the people.

But Dorset is not only agricultural, it is also a land
of stone-quarries.   Next to farming, this is its principal

A Dorset Farm, West Lulworth

industry, the great centres being Portland and the Isle of
Purbeck.   The geological structure of the county, with
its Portland and Purbeck stones and its Kimmeridge clay,
has chiefly contributed to this special industry, just as the
purity of the local water is supposed to contribute to the
excellence of Dorset ales.   Besides these activities, there
are much brick-making, sail-cloth and net-making, rope-

making, and large torpedo-works at Wyke Regis. Further sources of income to the county are its attractive watering-places, of which Weymouth is queen. Swanage, on the east, is overshadowed by the more fashionable Bournemouth, which lies just within the borders of Hampshire; but the whole Poole district gains by the proximity of its popular neighbour, and Parkstone, lying in Dorset, is practically a suburb of Bournemouth. The charming little Lyme Regis, at the western extremity of the coast, may suffer a little from the near competition of Devonshire, whose south-eastern watering-places are in general more accessible, but it retains and adds to the number of its lovers. Charmouth also attracts some, and West Bay is not without its popularity. Weymouth, besides its charm as a seaside resort and its excellence as a centre for excursions, is important as being the Great Western Railway port for the Channel Isles. There are other delightful places along the coast, more fully dealt with elsewhere, which draw large numbers of tourists.

Dorset is not a populous county. There is no single town that can be termed large, and Poole and Weymouth, the two largest, do not number even half the population of Bath or Exeter. In country districts the life is leisurely and placid; perhaps as Arcadian as anything can be in these strenuous times. Yet the placidity is not stagnation. Modernism—especially the modernism of the average villa-builder—is creeping in to mar the beauty of the thatched cottages, to disturb the "haunts of ancient peace"; but there is still much left of a rural loveliness that no other county can excel.

3. Size. Shape. Boundaries.

Dorset cannot claim to be one of the largest of English counties, though it is one of the most complete in the variety of its physical features. It embraces an extent of 625,578 acres, or in other words a little less than 1000 square miles. Of the forty English counties there are twenty-two larger and seventeen smaller. The counties that are nearest to it in acreage are Northampton, 642,393 acres; Warwickshire, 626,364; and Nottinghamshire, 616,287. These are all counties of medium size; we may compare them with the extremes—Yorkshire, 3,895,680 acres; Lincolnshire, 1,659,647; Devon, 1,633,269, these being the largest English counties; and Middlesex, 178,606 acres; Rutland, 108,700, the two smallest.

The shape of Dorset is roughly a parallelogram, and it lies with tolerable exactitude E., W., S., and N. Its greatest length, N.E. to S.W., 54 miles, is from a spot close to Gillingham station to the borders of Devon, about a mile westward of Lyme; its greatest breadth, 40 miles, is from the parish of Zeals to Portland Bill. Its borders march with four counties, Hampshire on the E., Wiltshire on the N.E., Somerset on the N.W., and Devon on the W. Southward is the English Channel. It should be noted that Dorset formerly extended westward to include Hawkchurch, Tytherleigh, and Chardstock, but this corner was transferred to Devon in 1896. The change may be seen by reference to earlier maps. As

against this loss the county has gained Thorncombe, with
Ford Abbey, which used to belong to Devon. There
can be no doubt that the ancient boundary of the counties
here was the river Axe, but in modern times the frontier
has been arbitrary and somewhat shifting. Among other
gains have been those of Trent, Goathill, Sandford Orcas,
Poyntington, from Somerset; while Wambrook has been
transferred to that county. These latter changes took
place in 1896. The present delimitation of the county
seems almost purely artificial, but such was not always
the case; there was a great frontier of forests on the
north and north-east, of which Cranborne Chase is a
survival; on the east were the swampy flats of the Stour
and Avon; while on the west the rivers Yeo and Axe
assisted in forming a natural boundary.

The coast-line, 75 miles in length, is of special interest,
and geologically of great importance. It begins on the
east with the very broken and shallow inlet of Poole
Harbour, followed by the bays of Studland, Swanage, and
Durlston. The chief headlands here are the Foreland
and Ballard Point, the seaward bluff of a height that
reaches 584 feet. Below Swanage are Peverel Point and
Durlston Head; followed about four miles W. by St
Aldhelm's Head, wrongly named St Alban's by those
who knew more of Alban than of the local but very
great Aldhelm. Then follows the Kimmeridge Ledge,
giving its name (that of a neighbouring village) to the
famed Kimmeridge Clay. This corner of the shire, from
Poole Bay to Worbarrow Bay, is known, inaccurately, as
the Isle of Purbeck. Bounded by the sea, the waters of

Poole Harbour, and on the N.W. and W. by the river
Frome and its tributary Luckworth Lake, it is nearly
insular, but has never been quite so.   Purbeck Isle is
about twelve miles in length and ten in breadth.   The
bays here are small till we reach Weymouth and Portland,
where the coastal configuration lends itself to the construction of the finest artificial harbour in the kingdom.

Typical Dorset Coast, Durdle Bay

The noted Chesil Bank, about eleven miles in length,
connects Portland Isle with the mainland at Abbotsbury ;
and from thence to Lyme Regis the coast-line is almost
unbroken.   This is a region of landslips, but these and
other features of the coast will be dealt with more fully
in the two Sections " Round the Coast."

As already mentioned, the flats of the Stour and Avon

form a strongly marked boundary on the east, where the county marches with Hampshire. This part belonged to the Andredsweald of Saxon times, and, though now largely denuded of trees, was then thickly wooded, being in fact a portion of the immense forest of Selwood, of which Cranborne Chase is a survival. Near Cranborne the county touches the Wiltshire border, which it pursues past Cranborne Chase to Shaftesbury, the shire reaching its highest point by Zeals, not far from the source of the Stour. There is then a sharp descent westward, and near Templecombe the border of Somerset is met. A succession of small county towns stand on the Somerset side of the boundary—Milborne Port, Yeovil, Crewkerne, Chard. Near Ford the Axe becomes the westward boundary of Dorset, with Axminster as in some sense a border town, in a district where, as mentioned, the delimitation has been very fluctuating.

4. Surface and General Features. Hill, Valley, and Heath.

Dorset has been spoken of as one of the most typical of southern shires, presenting in itself all the features that we chiefly associate with the counties of the south of England. These features are a varied coast-line, an interior of upland and valley, meadow, pasture and heath, river and woodland. But the characteristics of Dorset are those of the south-west rather than of the south-east; it is more akin to Devon, Somerset, and

Wiltshire than to Hampshire, Sussex, or Kent. Yet it shares a feature of Hampshire in its woodland (Cranborne Chase), and of Sussex in its chalk downs ; though with a difference. The distinction between the downs of Dorset and those of Wiltshire or Hampshire is that the former are more broken into ridges, more suggestive of primary

Typical Dorset Landscape, Shipton Hill, near Bridport

rock formation in appearance, while the downs of Wiltshire and Hampshire rise in rounded broad-heaving undulations. The average elevations are about the same, but are of course far less than those attained by the moorland ranges of Devon, or by the hills of Wales and of the Lake District. There is no eminence in Dorset reaching to a thousand feet, while the mean elevation of Dartmoor is about 1500 feet. Yet Dorset exhibits in little much of

what Devon exhibits in large; it is a county of great
resource, both in beauty and in productiveness.
Charles II once said that "he had never seen a finer
country, in England or out of it"; and Defoe thought
that a man might as well settle in Dorset as anywhere
else, if he wanted a desirable retreat. But there is great
dissimilarity in the features of the various districts. The
county, indeed, has been compared with Arabia—the
heathlands of the east being Dorset "Deserta"; the rich
western part "Felix," and the central "Petraea."

Geologically, on the other hand, the county belongs
to the middle-south and south-east rather than to the
west. The heart of the county is a vast upland, which is
attached in formation to the chalk of Salisbury Plain and
the Marlborough Downs, to the Chilterns, and to the
downs of Hampshire and Sussex. The north section of
the Dorset Downs extends in a southward and westward
direction from near Shaftesbury to the mouth of the Axe,
thus forming a kind of link between Wiltshire and Devon.
Their chief elevations are Melbury Hill, 863 feet; Oke-
ford Hill, 739 feet; Bell Hill, 846 feet; Bulbarrow,
902 feet; High Stoy, 860 feet; Lewesdon Hill, 894 feet;
and Pilsdon Pen, 909 feet; this latter being the highest
point in the county. The South Downs stretch from
near Beaminster in the west to Purbeck Isle and Poole
Bay; they are chiefly separated from the North Downs
by the valley of the Frome and the Great Heath. Somewhat lower in average height, their chief elevations are
Blackdown (or Blagdon), 789 feet; Eggardon Hill,
828 feet; and Nine Barrow Down, 655 feet. Most if

not all of these eminences are occupied by old camps, earthworks, and tumuli ; they were the home and the fighting-ground of departed races. Here Ivernian met Celt, and Celt met Saxon. The central core of upland is surrounded, on all sides but the west, by valleys of great luxuriance and fertility ; and the view from these Dorset heights, whether we look towards the sea or

Eggardon Hill

northward towards Somerset and Wiltshire, is always a wide panorama of rural opulence and beauty. Meadows and pastures, hedgerows, running waters, and little sequestered villages, are blended amid a wealth of scattered woodland, and with the fascination of lonely heathlands.

Perhaps the most characteristic, certainly the largest and most famous, of the Dorset valleys is the Vale of

Blackmore, of which the northern gateway belongs to Wiltshire. There was once a great forest of Blackmore, and there are still many fine trees surviving. It is like a fortunate island, an Arcadia, surrounded not by seas but by the protecting chalk-hills. Mr Thomas Hardy, whose novels have caused the modern world to regard Dorset as the very heart of Wessex—a distinction that historically belongs to Hampshire—says that the traveller who approaches the Vale of Blackmore from the coast is "surprised and delighted to behold, extended like a map beneath him, a country differing absolutely from that which he has passed through." The whole is a green chequer-work of pastures, only occasionally varied with grain-growing portions, while among the sparse villages there are grand clumps of oak trees, and the fertilising flow of pastoral rivulets. Well watered by the Lydden and the Cale, the Vale has a climate of soft humidity, without excessive rainfall; there is a dreamy blueness about the cradled air.

Very different from this is the Marshwood Vale of the far west, north-east of Lyme; this is both marshy and woody, and was described by the Dorset poet Crowe as "cold, vapourish, miry, wet." This, however, is only its winter character; in spring and early summer it is a perfect fairyland of flowers. Marshwood was noted (like Blackmore) for the fineness of its oaks, and for its great scarcity of stones, the smooth undulations yielding hardly a pebble. To the north rise Lewesdon Hill and Pilsdon Pen, two of the county's highest eminences; while the entrenched hills of Lambert's and Coney's Castles shut in the west. Southward are more hills, Stonebarrow and Golden Cap, and the Vale only opens towards the sea by the valley of the Brit.

Differing from these vales yet equally or more striking in its character, is the Great Heath, named in Domesday *Bruaria*. It stretches eastward of Dorchester to Poole Bay, and southward from Bere Regis to the Purbeck

The Heath Lands—Studland Heath

hills. The Roman road to Dorchester skirts its north-western border, and the highroad from Wareham to the same town is in part its southern boundary. It is watered by the Trent (or Piddle) and the Frome. Some would claim that it extends to Cranborne, but we may certainly take the Stour for its northward limit, the country lying beyond lacking the features that specially characterise the

Great Heath. The true Heath is an undulating stretch
of sand, covered with turf and heather, bracken and
gorse, with occasional bogs, mires, and ponds. The
passing seasons take it to themselves and bestow their
own features upon it, yet it retains an individuality of
aspect that is at times almost weird. Sir Frederick Treves
speaks of it as "a veritable part of that Britain the Celt
knew, since upon its untameable surface twenty centuries
have wrought no change. It is a primitive country still.
The wheat, the orchard trees, and the garden flowers on
its confines are products of civilisation, and are newcomers
to the land. Here, still living, are the rough, hardy
aborigines—the heather, the bracken, the gorse—which
settled on its heights when first they rose out of the sea."
In this primitive roughness and in its profusion of prehistoric remains the Heath resembles Dartmoor, but it is
a lowland, not an upland, and its formation is of sand
instead of granite. In old days great swamps reached
inland from Poole Harbour. Nature herself has reclaimed
much of this land; indeed, it would not need much change
to reclaim still larger parts of Poole Harbour. The town
of Wareham, surrounded still by its ancient earthworks
and entrenchments, may be regarded as the capital of the
Great Heath.

5. Rivers.

It cannot be said that the rivers of Dorset are of great
size or importance; but, deriving chiefly from the chalk,
they are clear and sparkling, and they do their work well

in watering the county, even if their navigable uses are slight. There is no river here so beautiful as the Wye or the Dart, the Tamar or the Fal; but for the most part the Dorset streams have a charm of simple pastoral loveliness, the charm attaching to the "running brooks" of a typical English landscape. They may also claim to be almost entirely local in origin and course. The main watershed is from the chalk-downs to the sea, and, as is not unusual, the streams intersect, rather than run parallel to, the ridges; yet the principal river, the Stour, rises just beyond the chalk. It springs from the Six Wells at Stourhead, about two miles within the boundary of Wiltshire, and though it reaches the sea at Christchurch in Hampshire, thus ending, as it begins, outside the county, nearly 60 of its 65 miles belong solely to Dorset. It gives its name liberally to towns and villages along its course—the Sturminsters, the Stowers, Stourpaine, and others; but we may note that this name, Stour, is shared by five or six other rivers in the kingdom, and though its etymology is not certain, it probably derives from a root meaning water. Its principal tributaries are the Cale, which enters the shire from Somerset; the Lydden; the Tarrant, which gives its name to some interesting parishes; and the Allen, or Win, from which are named Wimborne St Giles and the more famous Wimborne Minster. The Allen rises in Cranborne Chase. The Stour has another feeder in the Winterborne, which rises from the chalk about five miles west of Blandford, lending its name to a number of villages on its way; but as its name implies (winter-burn or brook) it only flows during the winter months, degenerating in summer to a mere trickle, or more often drying entirely. It is supposed that this drying is due " to the cessation of a syphon-action in connection with reservoirs of water in the downs." There is another brook of similar name and characteristics to the south-west of Dorchester, and the two between them give name to no less than fourteen

The Stour and Canford Bridge at Wimborne

Winterbornes in Dorset. If a stream can claim distinction from the number of places named after it, then the Piddle is almost as notable as the Stour; but, unfortunately for this claim, the course of the Piddle (now often called the Trent) is only about 25 miles in length. Rising in Alton Pancras, to the west of Nettlecomb Tout, we can trace it by the villages named after it along its course.

The little stream eventually falls into Poole Harbour near Wareham.

The second largest river of Dorset, the Frome, also flows into Poole Harbour, very near the Trent, after a course of 35 miles. It rises on the northern slope of the downs near Corscombe, receiving the Hooke at Maiden Newton and the Cerne (or Churn) near Dorchester.

The Frome—Maiden Newton Mill

Leaving the county-town behind it, the Frome zigzags past West Stafford and Wool to Wareham, giving its name to several Fromes and to Frampton. The Cerne, its chief tributary, comes from High Stoy, one of Dorset's highest hills, and passes the romantic Cerne Abbas on its course. It should be noted that all these rivers, the Stour, the Trent, the Frome, and also the Corfe on

Purbeck Isle, drain eastward to the Great Heath. The remaining rivers, the Wey of Weymouth, the Brit of Bridport, the Bredy of Burton Bradstock, the Char of Charmouth, take a more or less southern or westward course, and flow into the English Channel. The Axe, though it rises in Dorset and is the natural western boundary of the shire, must now be regarded as a Devonshire river ; just as the Yeo and the Parrett, also rising in Dorset, belong properly to Somerset.

There is nothing that can be called a lake in Dorset, unless we except the lake at Sherborne Park ; but there are some pleasant little sheets of water, which may be termed pools or ponds. Poole Harbour, at least in appearance, is somewhat akin to the Broads of East Anglia. The Fleet, to the west of Portland, is really a long narrow inlet of the sea. It is about nine miles in length, varying in breadth from a quarter-of-a-mile to nearly a mile, and is divided from the sea by the unique Chesil Bank, a natural breakwater. There has been much controversy as to the origin of this bank, which is the most singular of its kind in Europe. It extends far west of the Fleet, beginning near Burton Cliff as fine gravel which increases to pebbles of about three inches in diameter. There is something a little like it, though on a much smaller scale, at Slapton Lea in Devon.

Chesil Bank from Portland

6. Geology and Soil.

By Geology we mean the study of the rocks, and we must at the outset explain that the term *rock* is used by the geologist without any reference to the hardness or compactness of the material to which the name is applied; thus he speaks of loose sand as a rock equally with a hard substance like granite.

Rocks are of two kinds, (1) those laid down mostly under water, (2) those due to the action of fire.

The first kind may be compared to sheets of paper one over the other. These sheets are called *beds*, and such beds are usually formed of sand (often containing pebbles), mud or clay, and limestone, or mixtures of these materials. They are laid down as flat or nearly flat sheets, but may afterwards be tilted as the result of movement of the earth's crust, just as one may tilt sheets of paper, folding them into arches and troughs, by pressing them at either end. Again, we may find the tops of the folds so produced wasted away as the result of the wearing action of rivers, glaciers, and sea-waves upon them, as one might cut off the tops of the folds of the paper with a pair of shears. This has happened with the ancient beds forming parts of the earth's crust, and we therefore often find them tilted, with the upper parts removed.

The other kinds of rocks are known as igneous rocks, which have been melted under the action of heat and become solid on cooling. When in the molten state they have been poured out at the surface as the lava of

	Names of Systems	Subdivisions	Characters of Rocks
TERTIARY	Recent Pleistocene	Metal Age Deposits Neolithic ,, Palaeolithic ,, Glacial ,,	Superficial Deposits
	Pliocene	Cromer Series Weybourne Crag Chillesford and Norwich Crags Red and Walton Crags Coralline Crag	Sands chiefly
	Miocene	Absent from Britain	
	Eocene	Fluviomarine Beds of Hampshire Bagshot Beds London Clay Oldhaven Beds, Woolwich and Reading Thanet Sands [Groups	Clays and Sands chiefly
SECONDARY	Cretaceous	Chalk Upper Greensand and Gault Lower Greensand Weald Clay Hastings Sands	Chalk at top Sandstones, Mud and Clays below
	Jurassic	Purbeck Beds Portland Beds Kimmeridge Clay Corallian Beds Oxford Clay and Kellaways Rock Cornbrash Forest Marble Great Oolite with Stonesfield Slate Inferior Oolite Lias—Upper, Middle, and Lower	Shales, Sandstones and Oolitic Limestones
	Triassic	Rhaetic Keuper Marls Keuper Sandstone Upper Bunter Sandstone Bunter Pebble Beds Lower Bunter Sandstone	Red Sandstones and Marls, Gypsum and Salt
PRIMARY	Permian	Magnesian Limestone and Sandstone Marl Slate Lower Permian Sandstone	Red Sandstones and Magnesian Limestone
	Carboniferous	Coal Measures Millstone Grit Mountain Limestone Basal Carboniferous Rocks	Sandstones, Shales and Coals at top Sandstones in middle Limestone and Shales below
	Devonian	Upper } Devonian and Old Red Sand- Mid } stone Lower }	Red Sandstones, Shales, Slates and Lime- stones
	Silurian	Ludlow Beds Wenlock Beds Llandovery Beds	Sandstones, Shales and Thin Limestones
	Ordovician	Caradoc Beds Llandeilo Beds Arenig Beds	Shales, Slates, Sandstones and Thin Limestones
	Cambrian	Tremadoc Slates Lingula Flags Menevian Beds Harlech Grits and Llanberis Slates	Slates and Sandstones
	Pre-Cambrian	No definite classification yet made	Sandstones, Slates and Volcanic Rocks

volcanoes, or have been forced into other rocks and cooled in the cracks and other places of weakness. Much material is also thrown out of volcanoes as volcanic ash and dust, and is piled up on the sides of the volcano. Such ashy material may be arranged in beds, so that it partakes to some extent of the qualities of the two great rock groups.

The production of beds is of great importance to geologists, for by means of these beds we can classify the rocks according to age. If we take two sheets of paper, and lay one on the top of the other on a table, the upper one has been laid down after the other. Similarly with two beds, the upper is also the newer, and the newer will remain on the top after earth-movements, save in very exceptional cases which need not be regarded by us here, and for general purposes we may regard any bed or set of beds resting on any other in our own country as being the newer bed or set.

The movements which affect beds may occur at different times. One set of beds may be laid down flat, then thrown into folds by movement, the tops of the beds worn off, and another set of beds laid down upon the worn surface of the older beds, the edges of which will abut against the oldest of the new set of flatly deposited beds, which latter may in turn undergo disturbance and renewal of their upper portions.

Again, after the formation of the beds many changes may occur in them. They may become hardened, pebble-beds being changed into conglomerates, sands into sand-stones, muds and clays into mudstones and shales, soft

Lulworth Cove, Stair Hole (showing contorted strata)

deposits of lime into limestone, and loose volcanic ashes into exceedingly hard rocks. They may also become cracked, and the cracks are often very regular, running in two directions at right angles one to the other. Such cracks are known as *joints*, and the joints are very important in affecting the physical geography of a district. Then, as the result of great pressure applied sideways, the rocks may be so changed that they can be split into thin slabs, which usually, though not necessarily, split along planes standing at high angles to the horizontal. Rocks affected in this way are known as *slates*.

If we could flatten out all the beds of England, and arrange them one over the other and bore a shaft through them, we should see them on the sides of the shaft, the newest appearing at the top and the oldest at the bottom. Such a shaft would have a depth of between 10,000 and 20,000 feet. The strata beds are divided into three great groups called Primary or Palaeozoic, Secondary or Mesozoic, and Tertiary or Cainozoic, and the lowest Primary rocks are the oldest rocks of Britain, which form as it were the foundation stones on which the other rocks rest. These may be spoken of as the Precambrian rocks. The three great groups are divided into minor divisions known as systems. The names of these systems are arranged in order in the table and the general characters of the rocks of each system are stated.

With these preliminary remarks we may now proceed to a brief account of the geology of the county.

The Dorset formations belong largely to the Oolitic (or Jurassic) system of the Secondary period. Oolite is a

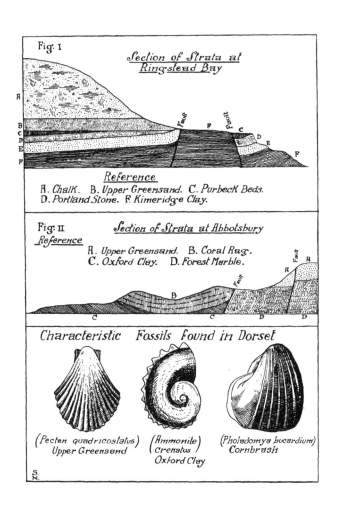

Fig. I

Section of Strata at
Ringstead Bay

R
B
C
D
E
F

Fault   F   Fault   C   D   E   F

Reference

R. *Chalk*.   B. *Upper Greensand*.   C. *Purbeck Beds*.
D. *Portland Stone*.   F. *Kimeridge Clay*.

Fig. II   *Section of Strata at Abbotsbury*
Reference

R. *Upper Greensand*.   B. *Coral Rag*.
C. *Oxford Clay*.   D. *Forest Marble*.

R   R

Fault

B

C   C   D   D

Characteristic   Fossils   Found in Dorset

(Pecten quadricostatus)   *(Ammonite)*   *(Pholadomya bucardium)*
Upper Greensand   crenatus   Cornbrash
Oxford Clay

S.
N.

species of granular limestone, divided into Lower, Middle, and Upper, according to its position, and nowhere else in England is Oolite better represented than in Dorsetshire. Speaking broadly, there is nothing in Dorset lower than the Lias (argillaceous or clayey limestone) that underlies the Oolite. This Lias is shown strongly at Lyme and eastward; it has yielded some fine specimens of fossil fish and saurians. One of these great lizard-like animals *Ichthyosaurus Platydon*, 25 feet long, was discovered a century since by a schoolgirl, Mary Anning, and is now at the South Kensington Museum. What is known as the Middle Lias may be seen between Bridport and Burton Bradstock, also running northward. It meets the Inferior Oolite, a subsection of the Lower, which also abounds in fossils and shells. One section of Lower Oolite, known as Fuller's Earth (from the use once made of it), consisting of marls and clays of blue and yellow, may be seen effectively at the coast near West Bay; and Forest Marble, the layer next above in position, may be found between Radipole and Abbotsbury and at Bothenhampton, often furnishing the stone used over a wide district as well as slates for building purposes. The Cornbrash, or highest layer of Lower Oolite, may be seen in the same district, and around Puncknowle and Weymouth; it begins in the north at Stalbridge. The Cornbrash is particularly rich in fossils. Of the Middle Oolite the chief is the Oxford Clay, found around Weymouth and Abbotsbury as well as in the north of the county, yielding curved oyster fossils. In the north, as also near Weymouth and Abbotsbury, we find Coral Rag, which yields excellent building material at Marnhull; a good section of this is exposed by the railroad at Sturminster Newton. Numerous Corallian fossils have been found near Sandsfoot Castle and at Osmington. At Abbotsbury the Corallian formation contains a large proportion of Limonite, or iron ore, its utility however being lessened by the presence of silica.

Devonshire Point, Lyme

But the great geologic pride of Dorset lies in the Upper Oolite, with its six divisions of Upper, Middle and Lower Purbeck, Portland Stone, Portland Sand, and Kimmeridge Clay. The Purbeck and Portland stones have gone to represent Dorset in London and many other distant towns. Purbeck marble is perhaps the most notable product of Dorset. It was used as far back as

Roman times in Silchester and other places, and in mediaeval it came to be much employed all over England, not only for the beautiful pillars that we can admire in so many of our cathedrals, but also for fonts, paving stones, effigies, and slabs for brasses. The Kimmeridge Clay takes its name from the village of Kimmeridge, between which and St Aldhelm's Head it is exhibited

Burning Cliff near Lyme Regis

at its best. In early British times it was worked into ornaments, very much as jet is still ; and circular discs of this material, often found in burial-urns and barrows, have been supposed to be a species of coin, popularly known as "coal-money." This is doubtful, but there is no doubt about the use of this clay as Kimmeridge coal, owing to the presence of bitumen. Cliffs themselves have been known to take fire, as happened at Ringstead Bay, near Lulworth, in 1826, when, ignited by spontaneous combustion, the cliffs continued burning for several years, sending forth volumes of stifling smoke and bluish flame. There is a burning cliff near Lyme Regis and a similar combustion is recorded as having occurred in the Lias cliff at Charmouth in 1751. Much of the Kimmeridge Clay however is not bituminous, but passes into a sandy grit. Among its fossils it has yielded parts of a fine Pliosaurus. Kimmeridge Clay is the main substratum of Portland Isle, where it underlies Portland Sand and Lower Purbeck. At St Aldhelm's Head the Portland sand and stone are well seen, overlying Kimmeridge Clay. The best Portland stone is known as the Whit or upper bed, which is extensively quarried on the Island. The Upper or Purbeck system of Oolite is itself divided into three series, the higher of which provides Purbeck Marble, a shelly limestone. Middle Purbeck is best seen between Kingston and Durlstone. The Great Dirt Bed of Purbeck is notable for its fossilised trees, found upright in their original position, with their native soil around them.

Leaving the Oolite we pass to the Lower Cretaceous, embracing the Hastings sand and Wealden clays, best exhibited between Worbarrow, Corfe, and Swanage. Nearly all the west Dorset hills have cappings of the Lower Cretaceous beds. The Upper Cretaceous gives us chalks and greensands, very widely extended; they enter the county at Shaftesbury, running southward and south-west to the borders of Devon, and then east by south to Swanage Bay. Chiefly consisting of Upper Greensand, this system also gives us a little Lower

Greensand, and Gault (clay and marl intersecting the greensands). The Eocene system gives us the famous Bagshot clays, well represented west of Poole and Wareham, and largely worked for ware and pottery. A curious feature in the Tertiary deposits is the hollowing of numerous pits, such as Culpepper's Dish and others in the Affpiddle district. This was once misinterpreted by archaeologists, who supposed that the hollows were the artificial site of prehistoric dwellings, but they are evidently the result of subsidence, or the removal of underlying chalk by some unknown action. Above the Bagshot clay we find a deposit that is probably Upper Pliocene in the Dewlish gravels, with traces that seem to speak of human operation ; and fossil remains of elephants have been found near the village of Dewlish. Of latest formation are the clays and high-level gravels of the surface—raised beaches, pebble beaches, and blown sands. These are more or less in a constant state of mutation. It may be noted that there are no traces of Glacial action in Dorset.

7. Natural History. Fauna and Flora.

In connection with its natural history we must remember that England, though an island, belongs strictly to the Continent of Europe to which it was once joined. Its insularity is, geologically speaking, comparatively recent. The effect of the Glacial Age, when much of our land was in the condition of Greenland at the present day, was to drive south or exterminate a very large proportion of the animals which inhabited it. When more favourable conditions returned the restocking took place from the east and south. It is always dangerous to mention dates or the probable duration of periods in dealing with geology ; but it is certain that a long time must have elapsed after the passing of the Glacial Age before the tides met around the shores of Britain and formed our islands. This insularising was almost certainly a gradual process, such as is still going on in parts of the eastern and southern coasts, rather than a sudden cataclysm Probably the climate and natural conditions of Britain were somewhat as they are to-day when the seas first covered the plateaux that are now submerged ; but sufficient time did not elapse after the passing away of the ice to allow all the Continental forms of life to make their way in from the south. Accordingly, England is somewhat poorer in these respects than the Continent, and Ireland with another and still deeper sea intervening is, as might be expected, still poorer than England. To take a familiar instance, forms of snake-life did not reach Ireland at all. It is probable that the last point of contact with the Continent was between Kent and the opposite mainland ; the shores of Dorset would have been washed by the sea long earlier.

Geology tells us something of what the natural history of Dorset once was in the far-off ages—how it included the gigantic ichthyosaurus, combining many of the characteristics of the lizards and the fish, sometimes measuring as much as 60 feet in length ; the plesiosaurus, almost equally immense, and the pliosaurus, which resembled the existing crocodile. Remains of these are

found in the Oxford and the Kimmeridge Clays of the
coastal regions. They inhabited the sea and estuaries
long before man had appeared on the earth, during what
is known as the Secondary period of geology ; but it is
certain that man existed in Dorset side by side with the
rhinoceros and elephant, the cave bear, the bison, the
hyena and the reindeer. Feeble though he was, he was
yet able to exist and multiply by the force of reason
which taught him to arm and protect himself, and to
outwit the monsters that surrounded him. The central

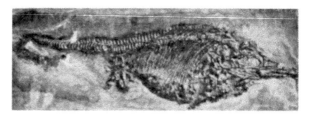

Fossil of Lyme Regis

figure among these formidable creatures, says Professor
Dawkins, was man : "a river-drift hunter who held his
own in the terrible struggle for existence." But even
the least formidable among these wild beasts and the
longest to survive—the boar, the wolf, the bear—have
been extinct for many centuries, and there is scarcely
anything that can be termed a wild beast in Dorset except
the roe-deer of the woodlands and the diminishing badger.

The animals of a district, like its plants,—or, it might
be more correctly said, with its plants, for there is a more
or less intimate connection between the two—depend not

a little upon the geological conditions. We may say, for example, that the great renown of Dorset butter is only the commercial expression of the fact that there is a great preponderance of the Oxford and Kimmeridge Clays, while the numerous sheep are the necessary outcome of the chalk uplands. The rule is still better exemplified in the case of the wild animals and birds. The heathlands, for example, which form so marked and peculiar a feature of the county have their own fauna. Here, a hundred years ago, black game were common, and are still found, though only on the preserved ground. Here too are the favoured haunts of the pochard and sheldrake.

If we look at the position of Dorset on the map of England we note that it is a maritime county on the English Channel about midway between the Land's End on the one hand and the S. Foreland on the other. We should expect, therefore, to find that so far as concerns its bird life it is commonly visited by stragglers from the south. This is the case, and the hoopoe, golden oriole, lesser bustard, and various other interesting species have often been recorded, but it is not so much in their path as the coast further to the east, and hence Sussex and Kent are richer in this respect. Yet it is not a truly western county, despite the fact that the chough, now unknown, used to breed in its cliffs. Still, however, an occasional American straggler finds its way to the county, though here again they are much fewer than in Devon and incomparably fewer than in Cornwall. The wild heathlands not only afforded suitable ground for black game but also for such birds as the Dartford warbler, but Mr Mansell Pleydell records that the winter of 1886–7 killed them almost if not entirely off. The so-called bearded tit is in our country essentially the bird of the Norfolk Broads, but it has also been recorded from the Poole district, where it presumably found similar conditions. The fine cliffs of Dorsetshire still afford a breeding-place for the peregrine falcon, especially about Purbeck, where it is protected; and the raven is not extinct. On the chalk downs in summer the thick-knee or Norfolk plover finds the special conditions it loves; while among the pebbles of the Chesil Bank the ringed plover and common tern have not less suitable ground, and nest abundantly. Poole Harbour is in winter a noted resort of wild fowl.

As we must take the origin of the name of Swanage to be Swan-wich, it is clear that the district was early notable for its swans; and no account of the natural history of Dorset would be complete without a reference to the great swannery at Abbotsbury. Considerably more than a thousand swans may be seen in this oozy region of creeks and pools, and the spot is haunted by wild-fowl. There are decoys here, as on the Broads of eastern England, for the capture of these latter, and similar traps are used on the Decoy Heath, a part of the Great Heath near Poole Harbour. The Abbotsbury swannery once belonged to the monastery of that place.

At Hook common, under Eggardon, it is said that no less than 60 species of our British butterflies have been taken, and the county is rich in this respect. The

Lulworth skipper, though not confined to Dorset, is a special feature of it.

Coming to the flora, we find that the ferns of Dorset deserve special notice ; among the rarer kinds being *Polypodium phegopteris*, found in Heffleton woods: *Asplenium marinum* or sea spleenwort, at Lulworth and Portland : *Ceterach officinarum* or scaly spleenwort, at Portland and

The Swannery, Abbotsbury

Wool : *Botrychium lunaria* or moonwort, which is rare : *Cystopteris montana* or mountain-fern, at Wool and Portland and also about Sherborne. The more common species are found freely in their natural localities, such as the hard fern, the prickly fern, the marsh fern (at Wool Heath), the maiden-hair spleenwort, the wall rue spleenwort, and the Osmunda or flowering fern.

Many plants of some rarity are to be found on Purbeck Isle and at Portland. Among these are the wild aster (*Aster tripolium*), several kinds of campanula and convolvulus, geranium, gentian, and iris; the *Rosa spinosissima* at Portland; *Scabiosa columbaria* on the Downs; three varieties of *Viola canina* at Portland; and though rarely, *Trifolium filiforme* and *Linaria elatine*.

Sherborne Park, Yew Tree Walk

Among the heaths may be found the exquisite *Erica cilia* which is peculiar to Dorset and Cornwall; while the bog asphodel, pimpernel, myrtle, marsh gentian, and buckbean are found freely on the Great Heath.

White and yellow water-lilies grow in the ditches, and sometimes the scarce bladderwort; walls give us eyebright, red bartsia, golden rod, etc.; river-banks have bulrushes and marigolds. The variety of the county's plants is greatly induced by the variety of its surface— chalk downs and fertile lowlands, heaths, bogs, and stony places. There are very fine oak-trees to be seen in the Blackmore Vale and in Lord Ilchester's park at Melbury ; there is a famous yew-tree walk in Sherborne park, elms flourish in the south-west, and the Abbotsbury neighbourhood has proved itself capable of raising sub-tropical vegetation, such as eucalyptus, olive, and other tender species.

8. Round the Coast. From Poole to Lyme Regis.

The coast of Dorset begins about two miles east of Poole Harbour, where it meets the sands of Bournemouth. The ancient port and the modern watering-place are now so intimately united by tramways and suburbs that they form an effective link between the counties of Hampshire and Dorset. Poole, the town and harbour, takes its name from the Celtic *pwl*, the Saxon *pôl*, signifying creek or inlet rather than our present sense of the word. The town has some doubtful claim to have been a Roman station, and is certainly of considerable antiquity. The harbour, entered by a very narrow mouth, is for the most part so shallow that it appears at low water like a section of the Great Heath ; but it is very beautiful when the tide is full, and has then some resemblance to the Broads of Norfolk. There is one island, Brownsea or Branksea, and several islets. It is well known that this part of the coast is affected by double tides, the earlier being the tidal wash from the Atlantic, the later that which enters the English Channel from the North Sea. At Poole this is further influenced by the peculiarities of its estuary. There is a flow of about six hours, followed by a brief ebb, thus causing, as it were, four tides in the twenty-four hours.

Brownsea Island

South of Poole Harbour are the sand-dunes of Studland Bay, leading to the chalk and greensand heights of Ballard Down, with its Foreland, Pinnacles, and Old Harry Point. There is a fine cavern here known as the Parson's Barn. The district from Poole to Worbarrow Bay is known, not very correctly, as the *Isle* of Purbeck, and has a finely varied coast-line and downs rising to about 600 feet in height. It consists of the famed Purbeck beds, Wealden,

Greensand, and Hastings Clays, underlain by Kimmeridge Clay. A couple of miles south of Old Harry Point is Swanage, most picturesquely situated and now a favourite watering-place. There is a good stretch of sand at Swanage Bay, leading southward to Peverel Point and Durlston. Passing the picturesque deserted quarries of Tilly Whim, we reach scenes of active quarrying at what

Mupe Bay

is called the Dancing Ledge, so named from the dance of the waves on a beach of solid stone. Near Winspit the East Indiaman *Halsewell* was wrecked in 1786, with the loss of 168 lives. St Aldhelm's Head (440 ft.), sometimes wrongly styled St Alban's, is named after the great Wessex bishop, and has an ancient chapel on its summit. Geologically, it exhibits Portland formations and Kimmeridge Clay. Beyond the little bay of Chapman's Pool is Encombe, with its small Freshwater cascade, and still further westward Kimmeridge Bay, Gad Cliff, and the beautiful bays of Mupe and Worbarrow, exhibiting a fine section of the Purbeck strata. The entrenched precipice of Ring's Hill here rises to a height of 560 feet, and there is a delightful verdant break in the chalk known as the

Arish Mell Gap and Cockpit

Arish Mell Gap. Just beyond the "Swine's Back" of Bindon Hill, with its fossilised trees, lies the remarkable cradled cove of Lulworth which attracts numberless tourists and sight-seers. It is entered by a narrow break in the cliffs, the sea having eaten away the chalk that once isolated the basin. Westward are some remarkably contorted strata. Westward of Durdle Bay and the Barn-door, a rocky archway haunted by sea-fowl, Swyre Head rises to 660 feet. At White Nore the coastward chalk ends, and at Ringstead are cliffs of Kimmeridge Clay which in 1826 became ignited and burnt for some years. The coast now sweeps boldly to the south to form Weymouth Bay, which is thus protected from north, south and westerly winds, but so exposed to the east that the town was severely damaged, and its esplanade entirely destroyed, during a great storm of the year 1824. Weymouth, anciently consisting of two boroughs, Weymouth and Melcombe Regis, has, like Poole, a tidal peculiarity in a secondary flow following the ebb, specially noticeable during Spring tides ; and the sailors incorrectly speak of " four hours' flow, four hours' ebb, four hours' standing water." Immediately below Weymouth is Portland Harbour, and the so-called Isle of Portland, which just escapes being actually insular by reason of the blown sand and pebble ridge extending from the Chesil Bank. Owing to its fine breakwaters, the harbour is now among the largest and best in the world. The Isle, chiefly consisting of Lower Purbeck and Portland Stone, superposed on Kimmeridge Clay, terminates with the oolitic quarried mass of Portland Bill, between which and some outlying banks the tide rushes with great fierceness, forming what is known as the Race of Portland. There is a raised beach at Portland Bill, composed of stones, pebbles, gravel, and sand, consolidated by water-action into a natural concrete ; this reaches to 65 feet above sea-level, and points to a considerable elevation of the land. Its original consistence may have been similar to

that of the Chesil Bank. To all appearance Melcombe
Regis itself rests on a kind of raised beach, while the
valley slopes of Preston, a mile or so eastward, yield
marine shells, signifying a great displacement in land and
sea-levels. Near Upwey there are terraces, known as
"Linchets," which have been explained as raised sea-
beaches.

Portland Bill

The coast-line from Chesilton to Burton cliffs is
practically unbroken. It is at West Bay, the mouth of
the Brit, that the remarkable formation termed the Chesil
Bank really commences, though it is usual only to apply that
name to the portion that separates the Fleet water from
the sea. Westward of the Brit the coast becomes more
beautiful, leading to the noble eminence of the Golden

Cap, 619 feet, and to Charmouth. The quaint little
port of Lyme Regis makes a fitting end to the county—
to which it has been so disloyal as to apply, vainly, for
admission into Devonshire. It combines the interest and
romance of its historic associations with extreme beauty
of surroundings.

9. Round the Coast. Sea-Encroachments, Landslips, etc.

The coast of Dorset, like the whole of the south
coast eastward of Tor Bay, consists largely of rock-
measures that are not strongly resistant to sea action ;
and though Dorset has suffered less in this respect than
Sussex and Kent, or than the Eastern counties, it has still
had to pay its price to the forces that made England
insular. Poole Harbour has all the signs of having been
shaped by comparatively recent water action. In a sense
it is the estuary of two rivers, whose channels keep it
clear ; but it also seems to be the remains of a much
larger Broad or marsh, which formerly covered part of
what is now the Great Heath and probably reached near
to the walls of Wareham. The ease with which the
Danes reached Wareham seems to prove that it was more
accessible by water in the past than it is now. At the
present time this miniature Zuyder Zee might be drained
without great trouble. A certain amount of erosion
takes place around the Purbeck peninsula, but there is
nothing notable to record till we reach Weymouth.
Sandsfoot Castle, now close to the sea, was once standing

in a field, surrounded by a moat; and Weymouth itself
has been severely devastated by storms.

Portland Isle has experienced several severe landslips.
One occurred on the north-east in the year 1615, when
100 yards of earth slid into the sea and a pier was
destroyed. About 80 years later there was another
landslip, and a still more serious one in 1734, when

Sandsfoot Castle

150 yards of land sank into the sea, demolishing a road
and pier. More notable still was that of 1792, when the
land "from the top of the cliff to the waterside sank
50 feet perpendicular. The extent of ground thus moved
was one mile and a quarter north to south, and 600 yards
east to west." Beyond Portland runs the natural bulwark
of the Chesil Bank, a bolder and finer breakwater than

could be constructed by man, defending the coast against a sea that would otherwise be irresistible. Though beneficent to the land, it is merciless to the sailor, and has been the scene of numerous shipwrecks. In the memorable storm of 1824 even this great rampart proved unavailing; the sea burst over it, sweeping away the greater part of the village of East Fleet, and leaving only the chancel of its church. On two occasions the waves have played remarkable pranks with vessels, lifting one and depositing it unhurt on the top of the Chesil Bank, and carrying another clear over the ridge into the safe waters of Portland Harbour.

It is well known that Lyme lies in a region of land-slips—in fact, the town itself has an appearance of tumbling into the sea. In addition to sea-erosion there is the operation of land-springs causing a general tendency to slide. The cliffs between Lyme and Charmouth are wasting rapidly, it is said at the rate of three feet a year, and a lane that led between the two places has long since disappeared. Old inhabitants of the neighbourhood speak of a time when the Church Cliffs reached half a mile further seaward; this may be an exaggeration, but the whole coast here and westward towards Seaton is pre-carious, and nothing can be done to secure it. An extraordinary landslip happened in 1839 at Dowlands, when an area of 40 acres slipped seaward; but to visit this would take us into Devonshire.

Though the waters are comparatively shallow around the Dorset coast, they are very free from shoals or banks; what vessels have chiefly to fear is the shore itself. There are ridges of sand around Studland and Swanage Bays, and the entrance to Poole Harbour, with its double lights, is difficult.   There is another lighthouse on Anvil Point, of 2000 candle-power, close to Durlston Head and Tilly Whim.   An underlying shallow ridge, out of sight beneath St Aldhelm's Head, represents former denudations of the coast-line.

Anvil Lighthouse

South of Lulworth are some banks, but of no great importance.   The south breakwater of Portland has its lighthouse ; and there is a new powerful lighthouse, with a revolving light of 2500 candle-power, on Portland Bill, replacing its old High and Low Lights, warning seamen of the perilous Shambles and Portland Ledge.   The

Shambles, a dangerous bank about three miles S.E., is also indicated by its lightship. Portland Ledge is a much smaller bank, directly to the south of the Bill. There are other lighthouses at Weymouth and at Lyme.

10. Climate—Temperature, etc.

The climate of a country or district is, briefly, the average weather of that country or district, and it depends upon various factors, all mutually interacting—upon the latitude, the temperature, the direction and strength of the winds, the rainfall, the character of the soil, and the proximity of the district to the sea.

The differences in the climates of the world depend mainly upon latitude, but a scarcely less important factor is this proximity to the sea. Along any great climatic zone there will be found variations in proportion to this proximity, the extremes being "continental" climates in the centres of continents far from the oceans, and "insular" climates in small tracts surrounded by sea. Continental climates show great differences in seasonal temperatures, the winters tending to be unusually cold and the summers unusually warm, while the climate of insular tracts is characterised by equableness and also by greater dampness. Great Britain possesses, by reason of its position, a temperate insular climate, but its average annual temperature is much higher than could be expected from its latitude. The prevalent south-westerly winds cause a drift of the surface-waters of the Atlantic towards

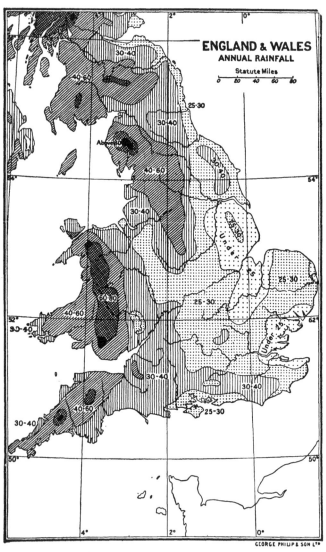

ENGLAND & WALES
ANNUAL RAINFALL

Statute Miles

(*The figures give the approximate annual rainfall in inches.*)

our shores, and this warm-water current, which we know
as the Gulf Stream, is the chief cause of the mildness of
our winters.

Most of our weather comes to us from the Atlantic.
It would be impossible here within the limits of a short
chapter to discuss fully the causes which affect or control
weather changes. It must suffice to say that the conditions
are in the main either cyclonic or anticyclonic, which
terms may be best explained, perhaps, by comparing the
air currents to a stream of water. In a stream a chain
of eddies may often be seen fringing the more steadily-
moving central water. Regarding the general north-
easterly moving air from the Atlantic as such a stream, a
chain of eddies may be developed in a belt parallel with
its general direction. This belt of eddies—or cyclones, as
they are termed—tends to shift its position, sometimes
passing over our islands, sometimes to the north or south
of them, and it is to this shifting that most of our weather
changes are due. Cyclonic conditions are associated with
a greater or less amount of atmospheric disturbance ;
anticyclonic with calms.

The prevalent Atlantic winds largely affect our island
in another way, namely in its rainfall. The air, heavily
laden with moisture from its passage over the ocean,
meets with elevated land-tracts directly it reaches our
shores—the moorland of Devon and Cornwall, the Welsh
mountains, or the fells of Cumberland and Westmorland
—and, blowing up the rising land-surface, parts with this
moisture as rain. To how great an extent this occurs is
best seen by reference to the accompanying map of the

annual rainfall of England, where it will at once be
noticed that the heaviest fall is in the west, and that it
decreases with remarkable regularity until the least fall
is reached on our eastern shores.   Thus in 1906, the
maximum rainfall for the year occurred at Glaslyn in the
Snowdon district, where 205 inches of rain fell; and the
lowest was at Boyton in Suffolk, with a record of just
under 20 inches.   These western highlands, therefore,
may not inaptly be compared to an umbrella, sheltering
the country further eastward from the rain.

The above causes, then, are those mainly concerned
in influencing the weather, but there are other and more
local factors which often affect greatly the climate of a
place, such, for example, as configuration, position, and
soil.   The shelter of a range of hills, a southern aspect,
a sandy soil, will thus produce conditions which may
differ greatly from those of a place—perhaps at no great
distance—situated on a wind-swept northern slope with
a cold clay soil.

The character of the climate of a country or district
influences, as everyone knows, both the cultivation of the
soil and the products which it yields, and thus indirectly
as well as directly, exercises a profound effect upon Man.
The banana-nourished dweller in a tropical island who
has but to tickle the earth with a hoe for it to laugh a
harvest is of different fibre morally and physically from
the inhabitant of northern climes who wins a scanty
subsistence from the land at the expense of unremitting
toil.   These are extremes; but even within the limits of
a county, perhaps, similar if smaller differences may be

noted, and the man of the plain or the valley is often distinct in type from his fellow of the hills.

Very minute records of the climate of our island are kept at numerous stations throughout the country, relating to the temperature, rainfall, force and direction of the wind, hours of sunshine, cloud conditions, and so forth, and are duly collected, tabulated, and averaged by the Meteorological Society. From these we are able to compare and contrast the climatic differences in various parts.

The climate of Dorset is preserved from extremes by the moderating influence of the sea, which is nowhere more than 40 miles distant. If it be admitted that the valleys are sometimes distinguished by mild humidity, it must be remembered that bracing down-lands or the sea-coast are never far distant. If we consult a map giving the isotherms (imaginary lines denoting mean temperatures), we shall find that, mainly by reason of its nearness to the west and south, Dorset lies within a comparatively warm zone. With the exception of the north-eastern part, and that portion of the coast lying east of St Aldhelm's Head, the mean annual temperature of the county is 51° Fahr. It will also be noted that the district south of Dorchester enjoys a slightly greater mild-ness during winter, and more coolness during summer. This district has also the largest annual average of sun-shine, while the mean sunshine of the whole county brings it within the more favoured region of Britain. At Weymouth the average yearly temperature is 50·8°, the same as that of Brighton ; but in sunshine Weymouth does better still, and a total of $1858\frac{1}{2}$ hours for the year was recorded. Dorset, indeed, always maintains an average that brings it within the most privileged district of British meteorology. In the matter of rainfall the shire occupies what we may call a medium position, with an average between 30 and 40 inches, though there are certain parts in the east, in the neighbourhood of St Aldhelm's Head and Poole Harbour and the coast near Weymouth, where the fall is between 25 and 30 inches. These districts lie too low to induce much condensation, while the Downs, where the highest fall takes place, have the usual high-ground effect of precipitating moisture.

In any consideration of climate we must recollect that local conditions cause variation, so that one parish will sometimes differ from its neighbour. Clay induces equality of temperature, while sands or gravels lend themselves to more rapid changes. The clay of Dorset plays its part in rendering the climate temperate and free from extremes. It does not necessarily follow that clay is best to live upon ; but the death-rate of Dorset proclaims it to be a healthy county, with an average of 14·1 as compared with 16·3 for the whole kingdom. The death-rate at Weymouth in 1904 was 11·2. The three chief watering-places, however, Weymouth, Swanage, and Lyme, all have the disadvantage of being exposed to the east.

11. People—Race, Dialect, etc.

Chiefly owing to the genius of one modern writer Dorset is commonly regarded as the very heart and type of old Wessex. But if the term be taken in its racial sense there is very little of the West Saxon in Dorset— perhaps less than in Somerset and parts of Devon. Nor can it be asserted that the people exhibit the common types of what we know as Celtic, though before the coming of the Saxons the county was occupied by tribes of Belgic Celts, the Durotriges and Morini. There was an earlier race whose stock largely survives. The first inhabitants of Britain were probably men belonging to the Palaeolithic or Old Stone Age, men whose stone weapons and implements were rough and unwrought, who came over dryshod before the land had become insular ; men, possibly, who were of similar race with the present Eskimo. They were succeeded at some unknown period by a race of men of the New Stone Age, of short stature, dark colouring, and long skulls, sometimes termed Iberians, Ivernians, or Silurians. These men had acquired far greater efficiency in the manufacture of their stone weapons and utensils, but did not yet work in metals, though they learnt this later. Colouring and the shape of the skull are physical features of great persistence, and judged by this standard the Iberian strain survives powerfully in the south-west and south-midlands of England, in Wales, and in Ireland. In these parts of the kingdom the older stock seems to have triumphed through all successive layers of

Palaeolithic Flint Implement
(*From Kent's Cavern, Torquay*)

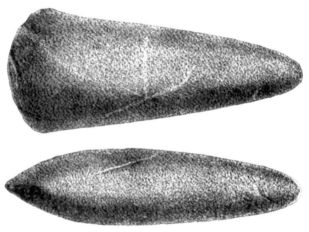

Neolithic Celt of Greenstone
(*From Bridlington, Yorks.*)

Celtic, Roman, and Teutonic immigration. The whole
of the south-west became little Saxonised in blood, but
the Saxons, when they at last arrived, were successful in
imposing their rule and their language on the Neolithic
peoples and their Belgic rulers who held such parts of
Dorset as were then populated. It must emphatically be
understood that language is no key to race ; the Welshman

Dolmen at Helstone

and the Irishman who speak English would be much
annoyed if we called them Saxons. The Wessex speech
therefore may survive strongly in Dorset, but the Dorset
folk are only in small measure West Saxons.

The Dorsetshire man is largely a blend, partly pre-
Celtic, partly Belgic (it has been disputed whether the
Belgae were Celts, but they are usually classified as such),
with some infusion of Teutonic blood. The average is

of middle height, with dark hair; the men are often broad
and strongly built, the women well-shaped and comely.
In the country places—and Dorset is almost entirely
country—they are leisurely, conservative in their notions,
shrewd, humorous, and kindly; preserving the characteristic features of English peasantry comparatively unspoiled,
by reason of their general isolation from railways. But
in all these matters changes have come and are coming.
The old "Do'set" dialect, seldom heard now in the
towns, lingers in the villages; many of the older folk
speak it to perfection. It is a true development of the
old Wessex speech, little corrupted; though of course a
West Saxon of a thousand years ago would only in part
have understood it. We must regard it as rather the
speech of a district than of a county, for it prevails, with
local variations, in Dorset, Wiltshire, Somerset, and Devon
—the chief Devonshire peculiarity being in the vowel-
sounds. There is a prevalent use of double-vowels, as in
"geate" for gate; of *d* for the thick *th*, as "drough"
for through; and of *v* for *f*. William Barnes, one of the
best authorities on the subject and one of the best users
of it for literary purposes, says : "This dialect, which is
purer and more regular than that which has been adopted
as the national speech, is, I think, with little variation
that of most of those western parts of England which
were included in the kingdom of the West Saxons, and
has come down by independent descent from the Saxon
dialect which our forefathers brought from the South of
Denmark, and the Saxon islands Nordstrand, Busen, and
Heligoland." He also stated that if the court had not

been moved to London, the Wessex speech, "the speech of King Alfred," would have been the prevailing language throughout England to-day. But to say this is to ignore the enrichment brought by Norman-French; and though in a sentimental way we may regret the decay of the old Wessex dialect, we cannot be otherwise than thankful for the growth, borrowings, and additions that have given us the language of Shakespeare and Milton, of Wordsworth and Addison. The languages that are philologically purest lack the wealth and resource of those that are hybrid.

The population of Dorset at the last census (1901) was 202,962, being about 207 to the square mile; this was an increase of 8,400 over the figures of 1891, and we may expect at least a similar increase for 1911. Perhaps the undoubted growth of Weymouth, Portland, Swanage, Parkstone, and Branksome, will give a still greater increase; but it is likely that the agricultural centres will prove stationary, even if they show no decline. We cannot expect Dorset to show such a population per square mile as the average for England and Wales, which is 558; but it must be remembered that agricultural labours employ comparatively few hands, especially in these days of machinery. The quarrying and stone-works of Portland and Swanage have brought a thicker population to those parts than we find in even the most fertile valleys of pasture and arable; while districts such as the Great Heath are almost entirely unpeopled. The figures show that females number $2\frac{1}{2}$ per cent. more than males, which seems to prove that the men stay at home more in Dorset than in some parts, where the excess of women is far

larger. Comparing Dorset with other counties, we find
that Warwickshire, of about the same acreage, has nearly
four times the population ; Nottinghamshire, a little
smaller, has considerably more than double the population ; Derbyshire, smaller still, has much more than
double ; Gloucestershire, rather larger, has about three
times the population. Lincolnshire, nearly thrice the
size, has more than double the population, and is therefore much more thinly peopled. When we recollect that
Dorset has no town of more than 20,000 inhabitants, it
will be seen that its population is fairly satisfactory.

12. Agriculture.

Dorset is not an industrial county in the north-country
sense of the term ; it does not depend on factories or the
making-up of imported raw materials, nor on the working
of steel and iron. Its atmosphere and occupations, in the
main, are agricultural and pastoral, the life of the farmer,
the field-labourer, the shepherd. Compared with the
hurry and turmoil of the great factory towns—towns
whose population in many cases exceeds that of this
entire county—the life seems peaceful and leisurely ; but
the law of compensation provides for loss as well as gain,
and if the city is apt to be feverish, restless, superficial,
excitable, the country has also its danger of becoming
stolid, stagnant, impervious to new ideas, and perhaps
intolerant. Those who desire to study the character of
the Dorset labourer cannot do better than turn to the

books of Mr Thomas Hardy and the poems of William
Barnes. These are the work of two men who have both
loved their native county dearly. The people of Dorset
in a double sense live on their soil ; they are supported
by its produce, by the pasture that it provides for their
flocks, and by the valuable stone which they raise from
its quarries. This business of quarrying is the only occu-

Butter Making on Cokers Frome Dairy

pation that can in any way compete with the shire's
agriculture.

The total acreage of Dorset being 625,578, we
find that as much as 479,744 acres lie under crops
and grass ; these figures, for the year 1907, showing a
slight increase over 1906. We shall realise what an
excellent proportion this is if we compare it with Hampshire, which cultivates 697,654 acres out of a total area

of 1,048,808; or with Cumberland, which cultivates 573,232 out of 961,544. The proportion, however, is beaten by the two neighbouring counties of Somerset and Wiltshire. A further examination shows us that, of these 479,744 acres, 169,121 are arable, and the remainder is under permanent grass. Wheat, with 17,818 acres, marks a decline on 1906; its quality is high, being nurtured by the fertile marlstones of the Brit valley and the clayey loams of the southern lowlands. Barley and rye have also diminished, but there is an increase in oats, the figure being 33,444 acres in 1907 against 31,311 for the preceding year. The total for corn-crops is a decrease of nearly 2000 acres, and any diminution of the country's power to feed itself is always regrettable. Dorset grows less corn than its neighbour Devonshire, and still less can it compare with such eastern counties as Essex and Norfolk. The figures for Essex are 508,845 acres arable, 284,101 under grass; while Norfolk gives 781,094 arable against 286,585 grass.

About 9 per cent. of Dorset's cultivated area is occupied by its owners; the remainder is divided into 4694 holdings, farmed for business purposes. The average size of the holdings is 95·8 acres, which we may compare with 60·8 in Somerset, 71 in Devon, 106·2 in Wiltshire, 71·7 in Cambridgeshire. As much as 27,868 acres of Dorset are returned as mountainous or heathland, but of course a large portion of this is available for grazing. Still larger is the extent under woodland or plantation, being 38,869 acres, a slight increase on previous years. Although the county is not particularly remarkable for its trees, there are fine oaks, yews, and elms at Cranborne Chase and elsewhere. Under orchard we have 4484 acres, almost the whole of which is devoted to apples ; the vegetable and small-fruit produce of Dorset is slight. There is a considerable culture of sainfoin.

Coming to the live stock of the county, we find that Dorset possesses 13,806 horses in agricultural use ; the

Broadmayne Sheep

total of horses returned for the shire being 16,590, being a decrease of sixty from the previous year. In this same year, 1907, the county possessed 90,333 head of cattle, a drop of more than 4000 ; there was also a diminution in the number of sheep, being 331,052 as against 334,605. But pigs showed an increase, 56,340 as compared with 51,424. The sheep include the famous long-horned "Dorset" breed ; while the short turf that covers the rocky soil around Portland supports a smaller breed, noted, like the Welsh, for the sweet juiciness of its mutton. The Dorset sheep are specially profitable on account of their early lambing.

It may be judged from these figures that Dorset well plays its part in the national economy. It deals in its native produce, turning its natural wealth to good account; imports and exports play a comparatively small part in its commerce. Excellent wheat and oats, good beef and mutton, fat pigs, world-famed butter—these are the things for which we look to Dorset. That county certainly does its work well which provides for the fit and whole-some feeding of the people ; and when it adds to this the providing of some of the finest building material in the kingdom, Dorset has no need to feel ashamed.

13. Industries. Quarrying, Fisheries.

As already mentioned, the prime occupation of Dorset is its agriculture, and it is this that keeps the population comparatively small, for an agricultural population is always thinly spread over a wide district. But the mineral wealth of the county, its geological formation in the coast-ward parts, provides a very good second line of industry by the quarrying of its Purbeck and Portland stones, and its clays. The county has sent much stone to the metropolis and to other large towns, and its supply gives no signs of exhaustion. The chief product is a species of limestone somewhat resembling the famous Bath stone but differing from it in being able to resist the eroding atmosphere of London. Portland stone was used sparingly in Exeter Cathedral about 1300, but it seems to have been re-discovered in Charles I's reign. Its merits were fully recognised by Sir Christopher Wren, who employed it in his building of St Paul's and other metropolitan erections. Inigo Jones used it for the Banqueting Hall

Portland Quarries

at Whitehall. The magnificent breakwaters of Portland are also constructed of this admirable material that lay so close to hand. The principal quarrying centre is at Portland, where the quarrymen form a distinct clan, with laws and customs of their own. The private quarries are of course quite separate from the Crown quarries, which are worked by the labour of the 700 convicts stationed here. Apart from convict labour, 1698 persons are engaged in quarrying and mining. Both the number of persons employed and the amount of output naturally vary from year to year; but, taking the latest available figures, we find that 243,680 tons of limestone have been raised in a single year, with 152,482 tons of clay, 49,931 tons of gravel and sand, 700 tons of chalk, and 40 tons of chert. These figures all apply to quarries

Swanage, Tilly Whim Caves

of more than 20 feet in depth; they might be increased by giving the entire output of all superficial quarrying throughout the county. Swanage is the centre of quarries which have been worked for a couple of centuries, like those of Portland. As shown by the above figures, an immense amount of very fine clay is also raised, the bulk of which goes to potteries in Staffordshire and abroad, though some is now held back for use in local potteries at Poole and Wareham.

Poole also does some shipbuilding, its yachts being specially excellent ; while other of its industries are cordage, sail-cloth, and shipping-requisites generally. Sail-cloth, sacking, nets, etc., are also produced at Bridport, Beaminster, Blandford, and other towns ; and Bridport was at one time the most important place of manufacture of flax and hempen cordage in the kingdom, and hangman's cords were known as " Bridport daggers." Its ropes and mats are still sent all over England. At Wyke Regis about 500 persons are engaged in the Whitehead Torpedo Works. Another Dorset industry is brewing, very excellent ale being produced at the Dorchester, Weymouth, and Gillingham breweries; while Gillingham also does some pottery and brick-work. Brick-making is fairly extensive in the county, from blends of sand and clay quarried at Broadmayne and elsewhere, and one result of agriculture is seen in large local flour-mills. Although the coast-line is so extensive, the fishing industry of Dorset is insignificant, and affords subsistence to very few persons. At Weymouth and Swanage the boatmen combine some fishing with their other pursuits, but they find summer visitors more lucrative than the precarious fish. Some fishing is done from Poole, Lulworth, Abbotsbury, Bridport, and Lyme, the chief catch being mackerel, but it is, in the main, little compared with the old days when, as Hutchins tells us, as many as 40,000 were sometimes taken at a draught, and 100 sold for a penny, though there are sometimes

exceptional catches. Though angling is now rather a
recreation than an industry, it may be mentioned there
is excellent fishing in the Frome and some other Dorset
rivers.

With regard to Portland Island, it is interesting to
notice that, stonework and fishing being the sole industries,
they are both often pursued by the same persons. When
quarrying is brisk, these men stay at home ; when it
becomes slack they take to their boats and follow fishery.
This dual occupation has been pursued by the same
families for centuries. There is a new industry developing
in Dorset which claims a few words, and that is lavender-
growing for the production of scent. A large tract of
land at Broadstone has been acquired by a firm of
cultivators, the greater portion of which is under lavender,
but thyme, peppermint, sweet balm, rosemary, and violets
are also cultivated with the same object. Two crops
of lavender are gathered, in midsummer and in early
autumn ; and the flower, packed in matting, is sent to
the distillery, where its essential oil, at an average of
$1\frac{1}{2}$ per cent., is extracted. The cottagers are not only
encouraged to grow scent-producing flowers in their own
gardens, but many have the additional employment of
weaving the dainty wicker baskets which are familiar as a
protection to the scent-bottles. This lavender industry is
nowhere else pursued so largely except perhaps at Hitchin.

14. Shipping and Trade. Seaports, Past and Present.

The chief glories of Dorset shipping lie in the past. At the present moment it can claim no seaport of prime importance—nothing to compete with Liverpool or Cardiff, Newcastle or Southampton. Yet it possesses, at

Weymouth

Portland, one of the finest roadsteads in the world, and the naval value of that position is likely to be fully appreciated in the future. If we wish to compare the past with the present, it is interesting to remember that Weymouth sent twenty ships to the siege of Calais, and Poole and Lyme four each, at a time when Liverpool could supply only one. Of these three Dorset ports two,

Poole and Weymouth, retain something of their importance, while Lyme has decayed into a picturesque fishing-
village and charming watering-place. Poole and Weymouth run a close race for supremacy in population, but
Weymouth is slightly the more populous town. Poole
however ranks first in shipping. Its seafaring annals are
varied and romantic, including a good deal of piracy and
privateering, and an extensive trade with Newfoundland,
still partially surviving. The quays of Poole are still
frequented by the beautiful old sailing vessels that are
becoming so scarce in our bustling modern ports, and an
old-world maritime atmosphere pervades the waterside.
The chief business is coast-wise ; and, taking the figures
for a single year, we find imports, of which the principal
is timber, to the value of £121,022, and exports, chiefly
clay and bricks, £13,192. But much of the clay goes off
by rail ; and there is also a considerable trade in sheep,
wool, butter, ale, etc. In one year 654 sailing and steam-
vessels of 95,332 tonnage entered the port with cargoes,
and 386 (30,176 tons) were cleared ; while 654 vessels
entered in ballast, and 860 were cleared. Weymouth,
which now ranks as second port, is chiefly concerned in
the trade with the Channel Islands, for which it is the
Great Western Railway point of departure. Vessels of
2000 tons can lie at its quay. Its imports in one year
totalled £585,897, of which the principal were potatoes,
tomatos, grapes, fresh flowers, and barley. The exports,
£363,898, are chiefly meat, grain, leather goods, cloth,
ale, etc. The Channel Isles are one of the kingdom's
great market-gardens, combining such produce with flori-

culture and vine-growing; it is to them, with some active competition from Scilly, that we look for much of our early potatoes, tomatos, and flowers. As regards many of the stores that are shipped to the Islands, Weymouth's share consists solely in transference of the goods from railway-truck to steamship. Weymouth, with the near proximity of Portland, is one of our few seaports that are adequately defended; its "Nothe" is strongly fortified.

Portland, now a first-class naval station, has a road-stead four miles long, protected by a fortified breakwater $1\frac{3}{4}$ miles in length. There is a powerful fort on the Verne hill, north of the island, and other batteries to the south. Much Portland stone and Portland cement are shipped from the docks.

The port of Lyme Regis, for custom-house purposes, belongs to Exeter. A considerable quantity of blue lias limestone is shipped, being used for making lime and cement. Bridport, whose harbour can only receive vessels of 250 tons, having silted-up as the estuaries in this part of the kingdom are apt to do, has a little coasting trade, and discharges a few timber vessels. As mentioned in the preceding Section, its local business is chiefly in the manufacture of rope, sail-cloth, nets, and the like; but its ancient fame for flax goods is still maintained by the presence of three large flax-mills. The chief export for the last thirty years or so, however, has been the agate sand, used for the paving-concrete of London. Millions of tons have been taken from the beach, with the result that the natural protection of the shore has been in great measure interfered with. Wareham, once more accessible, still has a small shipping activity, sending away quantities of potter's clay, pipe-clay, fire-clay, etc., and it also trades in lime and cement. The Dorset Cement Company gives employment to a number of persons, and the Sandford Pottery Company is engaged in stone-ware, pipe and tile-making. Swanage, occupied in straw-plaiting, the making of baskets and mats, etc., sends away some stone and clay by sea, but more by rail.

Lulworth Cove had once its small pier, but this has long since been destroyed. There are also at Kimmeridge Bay some relics of a demolished quay, speaking of some old disastrous attempts to make glass, and to produce burning oils from the Kimmeridge shale.

15. History of the County.

In past days the history of our present counties was largely determined by their natural position. Some of the existing shires are more or less arbitrary divisions; others, such as Dorset itself, were set apart and protected by natural boundaries. Thus Dorset was strongly shielded in the north and north-east by great forests (Selwood, Braden, and Cranborne); in the south-east it was guarded by the Great Heath and the morasses that surrounded Poole Harbour; southward it had the sea, and westward the river Axe. From the barrows, and some still older remains, we know that Dorset was peopled in very early times; and though history goes back no further than the Belgic Durotriges and Morini, we know that there must

have been at least two earlier occupations, in the Old
and New Stone Ages. The inhabitants immediately
preceding the Belgae were a people of dark skin and
hair (classified by Huxley as *Melanochroi*), whose strain
ultimately overpowered and survived the lighter skin and
fair hair of their invaders (*Xanthochroi*), and still mainly
prevails. The county was traversed by the track of the

Julian's Bridge, Wimborne

tin-merchants from Cornwall to Kent, whence the metal
was shipped to Gaul; the trackway entering near Lyme
Regis and passing out of Dorset in the north-east by the
existing road from Woodyates to Salisbury. This north-
east corner was strongly defended.

When the Romans came to England we have no record
of any severe fighting with them so far west as Dorset,

and the local tribes probably yielded to the inevitable without prolonged struggle. By the year 80 Agricola had succeeded in establishing the Roman peace throughout the whole of Britain ; the old trackways were converted into roads or "streets," the towns and strong places were garrisoned, and the wise Roman tried to win the people to admiration and emulation of his social and civil culture. The south, having long been amenable to Continental influences, was far less troublesome, and therefore less strongly garrisoned than the north. The county was connected by a branch of Icknield Street with the great network of Roman communications ; Dorchester, Lyme, Wimborne, and other towns were military stations ; villas were built with all the elegance and luxurious comforts of Roman taste ; and at Maumbury was a large amphitheatre for the sports in which Romans delighted.

But the Roman peace was not to endure. Hordes of Picts and Scots threatened it in the north, Teutonic vikings harassed it on the east and south-east coasts, and Rome herself was being stricken at the heart. It seems to have been about the year 436 that the last legion was withdrawn from our shores, and Britain, now largely Romanised, was left to her own internal dissensions and outward perils. When Hengist and his host landed on the Isle of Thanet in 449, the eastward Celts were doomed to a long period of storm and stress, but her natural defences gave Dorset a long respite. If Badbury Rings, as sometimes stated, was the scene of the great battle of Mons Badonicus, fought between the British forces and the West Saxons in the year 520, Dorset

can claim an important historic conflict; but the site
of this battle, in which King Arthur is said to have
led the Britons, is doubtful. If we could deliver
Arthur from the myths that encircle him, we should
probably find that he was a Romanised and Christianised
British chieftain. This fight of Mount Badon, if not
fought in Dorset, was at least fought near, and the county
enjoyed the reprieve that it afforded. It was not till the
beginning of the eighth century that Ine extended the
kingdom of Wessex to include Dorset; and the annexa-
tion was signalised by the founding of a bishopric at
Sherborne in 705, the famous Aldhelm being its first
bishop. Christianity by this time had taken some of the
sting out of the conquest.

A greater trouble was from the Danes. They began
to harass the Dorset coast in 787. Charmouth was a
special point of attack, as was Wareham, and both
suffered severely. In 857 the Anglo-Saxon Chronicle
tells us that "Ethelhelm the Ealdorman fought against
the Danish army at Portland isle with the men of Dorset,
and for a good while he put the enemy to flight; but the
Danish men had possession of the field, and slew the
ealdorman." Wareham was suffering continuously; it
was taken by the Danes in 876, while in the following
year the first great English naval fight took place off
Swanage, when Alfred, assisted by tempest, was victorious.
Dorset was much favoured by kings of Wessex, largely
for its hunting, and Ethelred, Alfred's predecessor, was
buried at Wimborne Minster, which was afterwards
destroyed by the Danes. There was a specially serious

descent and ravage by these Norsemen in 997. The first
landing of Canute, in 1015, took place at Wareham;
and he died at Shaftesbury in 1035. Shaftesbury Abbey,
founded by Alfred, had become a resort of pilgrimage
owing to the burial of Edward the Martyr, who was
treacherously murdered at Corfe by Elfrida, the mother
of Ethelred. A later interest attaches to Shaftesbury in

Swanage Beach

the recollection that the wife and daughter of Robert
Bruce were confined within its nunnery.

Wareham, said to have been the burying-place of
Beohtric, king of Wessex, had a castle erected by William
the Conqueror. This was held by Matilda in the time of
Stephen, but taken by the king and destroyed. Corfe was
also held for Matilda, by her faithful follower Baldwin de

Redvers ; and in this castle King John imprisoned Eleanor, the sister of Prince Arthur. Edward II was also captive here before his murder at Berkeley. Queen Margaret, landing at Weymouth on the very day of the disastrous battle of Barnet, found an asylum at Cerne Abbey with her son Edward ; and it was thence that she marched to the last fatal battle at Tewkesbury.

Dorset played a considerable part in the great Civil War, one specially notable event being the defence of Corfe Castle for the King in 1643, by the wife of Sir John Bankes. On the other hand, Lyme was held successfully for the Parliament against the utmost efforts of Prince Maurice. The siege lasted for about two months, and was relieved by the Earl of Essex just when famine was beginning to undermine the heroism of the people. Sherborne was besieged twice, and finally taken by Fairfax. Weymouth was held alternately for the King and the Parliament, and besieged by both ; and Portland suffered the same fate.

The county saw the last of the ill-fated " Clubmen," a gathering of West-countrymen, chiefly rustics armed with clubs or pitchforks, whose ostensible purpose was to defend their homes and fields from the damage done by the troops of both parties. Such was the sole aim of the majority, but the leaders and the secret working of the concern were undoubtedly Royalist. Hearing that a strong body of these were posted on the Castle Hill of Shaftesbury, Cromwell went forth from Sherborne against them, and by promises of just treatment induced many to return to their homes. About 2000, however, took

Corfe Castle

possession of the old British earthwork at Hambledon Hill, led by the rector of Compton Abbas. Cromwell sent what he calls a "forlorn hope of about fifty horse" to treat with them ; they fired and refused to listen. He sent a second time, saying that if they laid down their arms no harm should be done ; but "they still (through the animation of their leaders, and especially two vile ministers) refused ; I commanded your Captain-Lieutenant to draw up to them, to be in readiness to charge ; and if upon his falling-on they would lay down their arms, to accept and spare them. When we came near, they refused his offer, and let fly at him ; killed about two of his men, and at least four horses. The passage not being for above three abreast, kept us out : whereupon Major Desbrow wheeled about ; got in the rear of them, beat them from the work, and did some small execution upon them ;—I believe killed not twelve of them, but cut very many, and put them all to flight. We have taken about 300 ; many of which are poor silly creatures, whom if you please to let me send home, they promise to be very dutiful for time to come, and will be hanged before they come out again." Such is Cromwell's own account, which at least shows his proneness to mercy.

There is little more of historic consequence. Charmouth and Lyme had much to do with the escape of Charles II to France ; and Lyme was the scene of the landing of the Duke of Monmouth in 1685, rousing the people to enthusiasm on behalf of "the Protestant religion." It was an ill-doomed zeal, of which Judge Jeffreys made short work in his "Bloody Assize" after the battle of Sedgemoor. Poor Dorset afforded many victims to the merciless judge. Since that date the county has largely enjoyed the species of happiness that consists in " having no history."

Lyme Regis Parade

16. Antiquities—Prehistoric, Roman, etc. Earthworks, Camps, Barrows, etc.

In some ways no English county can surpass Dorset in the number and importance of its prehistoric remains. Among stone-circles, the Avebury and Stonehenge of Wiltshire are of course unmatched, but Dorset excels her neighbour in the size and number of her hill-fortresses, while in her sepulchral mounds and relics of Roman occupation she is at least her equal. Positively the grandest earthwork in Britain is that known as Maiden Castle (*mai-dun*, the "stronghold of the plain") near Dorchester, which is sometimes supposed to be the Dunium mentioned by Ptolemy. It is clearly the work of men of the later Stone Age, men who buried their dead in round barrows, and who raised this entrenchment, with merely their

Maiden Castle, Dorchester

stone picks or " celts " as tools, for a defence against such invaders as the Durotriges ; and the latter in their turn doubtless used it against the Romans—unless they submitted without fighting. Standing on an oval plateau that rises 400 feet above sea-level, the camp is defended to the north by triple ramparts, 60 feet high, to the south by five ramparts of somewhat less strength. The circumference of the whole is nearly two miles. At the chief entrances, east and west, the ramparts overlap with ingenious intricacy, and are increased in number. Considering the imperfect tools at their command, the builders of this earthwork produced a masterpiece of defence. Traces of Roman occupation have been discovered, proving that the Romans used it, as they also used the neighbouring camp of Poundbury, which enclosed about 20 acres. Both would have been used as military stations in connection with their settlement at Dorchester, and the Maumbury Rings, near by, was their huge amphitheatre, capable of accommodating about 12,000 persons and probably a site before Roman times. Badbury Rings, smaller than Maiden Castle but still a vast earthwork, with triple wall and fosse, is on the old prehistoric road from Salisbury to Dorchester; it has been identified, somewhat doubtfully, with the Mons Badonicus or Badon Hill where the Britons, said to have been led by King Arthur, defeated the West Saxons and for a while delayed their advance. There are round barrows near. On Pilsdon and Bulbarrow Heights there are other great earthworks, that of Bulbarrow known as Rawlsbury Rings; and there is a fine oval camp, resembling Maiden, at Eggardon. At Lewesdon near Sherborne is a very remarkable earthwork, and at Flowers Barrow near Lulworth, at Woodbury, Hod Hill, Hambledon Hill, and Weatherbury, will be found other of the 25 hill-fortresses enumerated by Hutchins as being earlier in construction than the Roman occupation. Shaftesbury is on the site of an early British town, the Caer Paladur of Geoffrey of Monmouth, and its Castle Hill has given traces of Roman presence in coins and masonry. The remarkable hive-shaped underground dwellings, found when quarrying at Portland and destroyed, were probably even earlier; and Dorset has plentiful traces of Iberian hut circles, of the Newer Stone Age.

Barrows or tumuli, chiefly round or bowl-shaped, are very common, with a few of the earlier long-barrows (at Blandford, Bere Regis, etc.) and some disc-shaped (at Long Bredy and Woodyates). At Rimbury, near Sutton Poyntz, a complete burial-ground was discovered, enclosing many urns and kistvaens (stone chests), with skeletons buried beneath the urns. This was clearly the cemetery of the ancient camp of Chalbury, close by. At Winterborne St Martin an immense barrow has been opened, disclosing many flint implements, pottery, knives, and a curious bronze dagger, besides skeletons. This barrow was probably associated with the neighbouring Maiden Castle, and is conjectured to date from about 2000 B.C. Though Dorset has no White Horse, it has its Cerne Giant, a gigantic figure, 180 feet in height, cut in the chalk of Giant's Hill, Cerne Abbas, once somewhat grass-grown but now kept better "scoured." It much resembles the Long Man of Wilmington in Sussex, and its date is equally uncertain. The county has some standing-stones or menhirs, but is not rich in this respect ; its earthworks are more notable than anything it has to show in stone.

When we come to antiquities that are definitely Roman, we find a very great wealth. Dorset was approached from

Old Sarum by the road known as Icknield Street, which
led by a direct south-western course to Dorchester, the
Roman *Durnovaria*. Though undoubtedly on the site of
an earlier settlement, Dorchester is distinctively a Roman
town, in spite of a present appearance that is perhaps
aggressively modern. As Mr Hardy says in one of his
books, "it was impossible to dig more than a foot or

The Giant, Cerne Abbas

two deep without coming upon some tall soldier of the
Empire, who had lain there in his silent unobtrusive rest
for 1500 years." Tesselated floors are so common in
Dorchester that the town did not scruple to present one,
not long since, to the younger Dorchester of the United
States. A specially fine one was discovered in 1899, of
highly decorated style ; it is now in the admirable County

Museum with many other Roman remains. Other beautiful tesselated pavings were unearthed in 1903, at Fifehead-Neville, near Sturminster Newton. A fine floor discovered at Frampton shows the Christian labarum (or monogram, ☧, of the Greek letters X and P, to denote the name of Christ) in combination with Neptune, Cupid, and other pagan emblems. The existing "Walks" of Dorchester follow the line of the Roman walls, of which a small trace survives, and the four intersecting streets are obviously of Roman pattern. Dorset was a part of the Roman Britannia Prima, and there were other stations at Wareham (*Morinio*, so called evidently from the Belgic tribe of Morini, whose great earthworks are the present walls of the town) ; at *Vindogladia*, of doubtful site, but probably either Wimborne or Woodyates ; at *Londinis*, Lyme; *Clavinium*, Jordan Hill, north-east of Weymouth ; and *Ibernio*, Bere Regis. *Canca Arixa*, Charmouth, and *Bolclanio*, Poole, are more dubious identifications ; and the debated *Moridunum* was almost certainly within the borders of Devon. There are also noteworthy " castra " at Cattistock, Hod Hill, Milborne, and other places ; and traces of Roman villas have been found at Rampisham, Sherborne, Halstock, Preston, etc. Many Roman remains have been disclosed on the Isle of Portland, the Roman *Vindilis*. The county has also yielded a few British coins and many Roman ones. Bokerly Dyke, a boundary entrenchment in the north-east of the county between Dorset and Wiltshire, was probably thrown up by the Durotriges, right across the old road from Sarum, after the departure of the Romans, as a defence of a rather

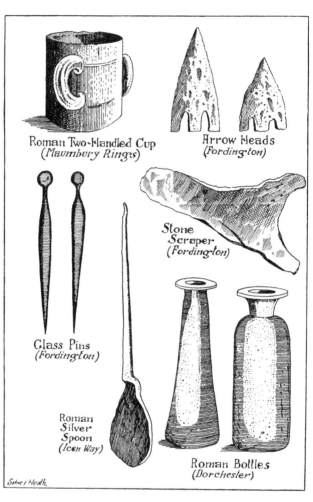

Roman Two-Handled Cup
(Maumbury Rings)

Arrow Heads
(Fordington)

Glass Pins
(Fordington)

Stone
Scraper
(Fordington)

Roman
Silver
Spoon
(Icen Way)

Roman Bottles
(Dorchester)

Sidney Heath.

Roman and other Antiquities in Dorchester Museum

vulnerable spot. This road and others in Dorset, though used and improved by the Romans, were certainly of far earlier date ; and originally they were doubtless mere tracks.

Although Poundbury Camp has been claimed for the Danes and may have been occupied by them, there are very few traces of the Danes in the county ; and Saxon remains are chiefly in connection with the old minsters. Perhaps a word should be given to the discovery at Fordington of an inscribed slab of Purbeck marble, pronounced to be of the first century, and supposed to be the tombstone of Aristobulus. An interesting, but doubtful tradition says that Aristobulus was a Roman disciple of St Peter and St Paul, and was sent to be bishop of the British.

17. Architecture—(*a*) Ecclesiastical.

Before speaking of the churches of Dorset it will be well to trace the leading features of the successive orders of ecclesiastical architecture. We do not usually speak of Roman architecture in connection with our churches, for there is no Roman church surviving in Britain ; but it is important to remember that the shape of our churches derives from the Roman basilica, which in some sense was the town-hall or guildhall of Roman times ; being a square building divided into three parts by two rows of pillars, these parts corresponding to the nave and aisles of a modern church. At the upper end was the tribunal, usually semicircular, resembling the apse of a chancel. The basilica, designed for secular and legislative purposes, was often converted to Christian use. In time the square was lengthened to an oblong, transepts were added, and the church took the form of a cross that is so common. The Celtic church of early times, of which traces survive in Cornwall, Wales, and Ireland, was generally an oblong of rugged masonry; the Saxons, less influenced by Roman example, did little work at first in stone, and their earlier churches were of wood. This working in wood left its definite influence on the stonework that was to follow.

For all practical purposes the orders of architecture with which we have to deal are the following, which, with the exception of Perpendicular, merged into each other by gradual transition, without sudden break, so that it is difficult to say where one ended and another began.

Pre-Norman or—as it is usually, though with no great certainty termed—Saxon building in England was the work of early craftsmen with an imperfect knowledge of stone construction, who commonly used rough rubble walls, no buttresses, small semicircular or triangular arches, and square towers with what is termed "long-and-short work" at the quoins or corners. It survives almost solely in portions of small churches.

The Norman Conquest started a widespread building of massive churches and castles in the continental style called Romanesque, which in England has got the name of "Norman." They had walls of great thickness, semi-circular vaults, round-headed doors and windows, and massive square towers.

Norman Font Chickerell

Outside Water Stoup
Broadmayne

Early English
Window
Wimborne

Norman Pier and
Arch-spring, Wimborne

Bell Turret
Radipole

S.
Heath

Perp. Capital
Upwey

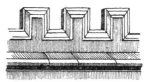

Perp. Parapet St. Peters
Dorchester

From 1150 to 1200 the building became lighter, the
arches pointed, and there was perfected the science of
vaulting, by which the weight is brought upon piers and
buttresses.   This method of building, the "Gothic,"
originated from the endeavour to cover the widest and
loftiest areas with the greatest economy of stone.   The
first English Gothic, called "Early English," from about
1180 to 1250, is characterised by slender piers (commonly
of marble), lofty pointed vaults, and long, narrow, lancet-
headed windows.   After 1250 the windows became
broader, divided up, and ornamented by patterns of
tracery, while in the vault the ribs were multiplied.   The
greatest elegance of English Gothic was reached from
1260 to 1290, at which date English sculpture was at
its highest, and art in painting, coloured glass making,
and general craftsmanship at its zenith.

After 1300 the structure of stone buildings began to
be overlaid with ornament, the window tracery and vault
ribs were of intricate patterns, the pinnacles and spires
loaded with crocket and ornament.   This later style is
known as "Decorated," and came to an end with the
Black Death, which stopped all building for a time.

With the changed conditions of life the type of
building changed.   With curious uniformity and quick-
ness the style called "Perpendicular"—which is unknown
abroad—developed after 1360 in all parts of England and
lasted with scarcely any change up to 1520.   As its name
implies, it is characterised by the perpendicular arrangement of the tracery and panels on walls and in windows,
and it is also distinguished by the flattened arches and the

Sherborne Abbey

square arrangement of the mouldings over them, by the elaborate vault-traceries (especially fan-vaulting), and by the use of flat roofs and towers without spires.

The mediaeval styles in England ended with the dissolution of the monasteries (1530–1540), for the Reformation checked the building of churches. There succeeded the building of manor-houses, in which the style called "Tudor" arose—distinguished by flat-headed windows, level ceilings, and panelled rooms. The ornaments of classic style were introduced under the influences of Renaissance sculpture and distinguish the "Jacobean" style, so called after James I. About this time the professional architect arose. Hitherto, building had been entirely in the hands of the builder and the craftsman.

Even in the case of our older churches, it is rare to find a building belonging to one unmixed order ; almost always there is a blend of several.

There was a time when Dorset was especially rich in monastic foundations, and abundant traces of this wealth survive in her churches. Foremost among these is the beautiful abbey of Sherborne, Saxon in origin, and retaining many features of Norman work, but transformed into Perpendicular by a great restoration in the fifteenth century. A blocked-up doorway and wall to the north replace the Saxon original. The south porch was Norman but together with many other parts has suffered from reconstruction. The tower is Norman to the floor of the bell-chamber. Bishop Roger's chapel and the Lady chapel are Early English. The choir is exquisite Perpendicular, with very fine fan-tracery ; the colouring,

Wimborne Minster

both of the interior and exterior, is rich and harmonious, the walls, of Hamhill stone, being a warm yellow. Part of the monastic buildings, the Guesten and Abbot's halls, Abbot's lodging and kitchen, survive, somewhat trans- formed, in the present grammar-school. As an abbey Sherborne belonged to the Benedictine order—the great rule of monachism established by St Benedict (born 480). Wimborne Minster, another fine church, also of rich external colouring, dates from the eighth century. It has a central Norman tower, above which was formerly a spire ; the western tower is Perpendicular. The nave is chiefly Norman and Decorated, and the choir Early English, with a striking east window of three lancets. Some of the tombs are specially notable, such as the altar-tomb of Beaufort, Duke of Somerset, and the Etricke tomb. The chained books in the library are widely famed.

Milton Abbey, established as a Benedictine foundation in 964, is so isolated that it may well be missed unless looked for. This beautiful minster, or rather portion of a minster, for it has no nave, was completed about 1320. The altar-screen is especially beautiful and the wooden tabernacle for the reservation of the Eucharist is a very rare example. Only fragments remain of Cerne, Shaftesbury, and Abbotsbury Abbeys, which were also Benedictine, or of Bindon, which was Cistercian. (The Cistercian rule, a branch of the Benedictine, was founded in 1098, in the French forest of Citeaux or *Cistercium*.) Shaftesbury was the most important of the Dorset Abbeys, one of the old Saxon nunneries peculiar to the south of

Chained Books, Wimborne

England which were allowed to continue rather than be refounded by the Conqueror's Abbots. The church is gone completely, but its foundations have been excavated. There is an old saying "Glaston and Shaston[1] owned more than the king." Ford Abbey, Cistercian, was in part built in 1148, but was much beautified shortly before the Dissolution, by Thomas Chard, its last abbot. It was converted into a private residence in the seventeenth century, the style being chiefly Tudor. The beautiful Chapter House is Transitional, from Norman to Early English.

Iwerne Minster is a blend of Norman, Early English, and Perpendicular, with a spired tower. Spires are very rare in Dorset; there are only two other specimens, one of which gives its name to Winterborne Steepleton; but the Perpendicular towers are usually fine, especially in the neighbourhood of Somerset, a very good example being that of Beaminster, with its rich sculpturing of the Tudor period. There is noteworthy Norman work at Worth Matravers, and an almost pure specimen at Studland. Specially admirable is the timber roof of the beautiful church at Bere Regis, with its projecting figures of the twelve Apostles, and its canopied Turberville tombs. There are interesting churches at Maiden Newton (Norman and Perpendicular), Wyke Regis (Perpendicular), Wareham, Piddletrenthide, Piddletown, Dorchester, Fordington (Norman, Early English, and Perpendicular), Bradford Abbas (Perpendicular), and Marnhull (chiefly Early English). It may be noted that

[1] The local name for Shaftesbury.

Ford Abbey

7—2

there is not a very great deal of Decorated in Dorset. But many more churches than those named possess features of beauty and interest, fine tombs and good old plate. It will usually be found that the parish church of the smallest village deserves a visit. On the whole, perhaps the most interesting church in Dorset is Whitchurch Canonicorum, where is the shrine of St Candida.

18. Architecture—(*b*) Military.

Dorsetshire is more notable for its surviving earthworks, pre-Celtic, Celtic, and Roman, than for what we understand in the modern sense as castles. Those that it once possessed have for the most part passed away, or have left only slight traces. But there is one glorious exception, the castle of Corfe, which even in its ruined state is one of the finest remains in the kingdom. Though we may guess that this gap in the Purbeck hills had long been fortified, there is no trace of anything earlier than a Saxon hunting-lodge, the scene of a striking event in English history when Aelfrith, or Elfrida, murdered King Edward the Martyr in order that her own son Ethelred might succeed to the throne (978). The name of Corfe (formerly *Corvesgate*) seems to be derived from the Anglo-Saxon *ceorfan*, to cut or carve, and doubtless refers to the cleavage in the hills which the castle dominates. Probably the Norman fortress was begun in the time of the Conqueror ; it is certain that a strong castle existed here in 1139, when King Stephen vainly besieged Baldwin de Redvers, who was holding it for Matilda. The cruel John used it as a residence, and also as a prison ; it is said that he caused some French prisoners to be starved to death here. At one time Eleanor, sister of Prince Arthur, was captive at Corfe, and also Edward II, afterwards murdered at Berkeley. After belonging to the

Corfe Castle

Somersets and to the Crown, Corfe passed to the Hattons; at one time it was an attached port, or "limb," as it was termed, to the Cinque Ports. It was bravely held by the wife of Sir John Bankes, while he fought with the King in the Civil War, and after a noble defence the siege was raised in August 1643. Three years later it was besieged again, and this time fell by the treachery of one of its garrison. It was condemned to be "slighted" or demolished by the Parliament, but was too strong for its destruction to be more than partial; and many of the massive ruins still lie just as they fell before the shock of Roundhead gunpowder. The walls follow the contour of the hill-summit, with a bridge leading from the first or outer ward across the ravine of the moat. Especially to be noticed are the great gateway, with its two side towers, the Plunkenet Tower and the Gloriette Bastion. The fine gateway of the second ward dates from the time of Edward I, and here we find that the powder of the would-be destroyers wrought something like a landslip, the western or Sunken Tower having been carried bodily about nine feet down the slope, yet still standing erect by reason of its solid masonry. Only two sides remain of the octagonal Buttavant (or *Bout avant*) Tower. Some traces of Saxon herring-bone work (flat stones or tiles placed bone-wise in rough masonry) were discovered during excavations here, and take us back in mind to the days of Edward the Martyr. The great keep, in majestic ruin, dates from the time of Henry I, and eastward of this is the Queen's Tower with its hall and chapel, raised on vaulted crypts. There is much else of interest; Corfe is a very noble and impressive ruin.

The remaining castles of Dorset are less important, but there are some striking remains at Sherborne. The present building, erected by Bishop Roger of Caen, once belonged to Walter Ralegh. It was successfully held for the King against the Earl of Bedford, but was taken in 1645 by the irresistible Fairfax, and then destroyed. Portions of the keep remain, and of a chapel with Norman windows; also the Norman gatehouse, with Tudor windows. Portland Castle, not specially interesting, was a block-house built by Henry VIII, and granted in turn to three of his wives. In the same district is the ruin of Sandsfoot, also built by Henry as a coast defence at a time when he feared that

Sandsfoot Castle and Portland Roads

the angry Pope might prompt an invasion of England; remains of the Norman and Early English abbey at Bindon were used in its construction. On the Isle of Portland itself is the Bow-and-Arrow, or Rufus, Castle, three hundred feet above sea-level, supposed to have been built by Rufus; beauty of situation is its best feature. Pennsylvania Castle, close by, is modern, having been built in 1800 at a cost of £20,000 by John Penn, Governor of Portland.

There was an old castle by the bridge of Sturminster Newton; the mound, now an orchard, still remains, but no trace of the original building. The ancient castle of West Lulworth has also disappeared; the existing Lulworth Castle is comparatively modern, dating from 1648. It is a fine square erection of solid masonry, with sturdy round towers at each angle. It has entertained many royal visitors, including James I, Charles II, George III, our present King Edward, and the refugee French King Charles X. The house contains some portraits by Lely.

Wolfeton House, at Charminster, a Tudor mansion with battlemented towers, dates from 1505, being enlarged a century later. In the year following its erection the Archduke Philip of Austria and his wife, being driven by tempest into Weymouth, were received here; and it was in connection with their visit that John Russell of Bridport, being summoned as interpreter, laid the foundations of that fortune which led to the establishment of the Dukedom of Bedford. It was a romantic instance of a sudden unexpected opportunity, utilised to the full. Woodsford Castle, supposed to have been built in the time of Edward III by Guy de Brian, retains only one of its four corner towers; it is now a farm, with little of the castle in appearance; but its vaulting, thick walls, chapel, and the immense old fireplace in its hall, tell of its past. The castles of Dorchester and Wareham are gone past trace.

Lulworth Castle

19. Architecture—(c) Domestic.   Seats, Manor Houses, etc.

Owing to the excellence of its native stone, it is natural to find Dorset rich in well-built houses; this even applies to its cottages, which are usually solidly substantial as well as picturesque.   Of the mansions and family seats,

Parnham House

some, being partly or wholly castellated (such as Woodsford Castle), have been mentioned in the preceding Section; but the county has some fine manors which can in no sense be termed castles.   Such, though now unfortunately much modernised, is the Tudor mansion of Parnham, at Beaminster, in its park of about 80 acres.   The hall, dating from 1449, bears heraldic devices of the Strode family on its glass and walls. To this family belonged William Strode, one of the "Five Members." The manor-house of Mapperton, beautifully situated in a glen about two miles distant, dates from Henry VIII; it is a two-storey building of yellow stone, with eagles on its gate-posts, beautiful bay and dormer windows, and

Athelhampton Hall

chimneys that look like pinnacles. Melplash House, in the same district, now a farm and much transformed, still speaks of the proud days when it was the home of Mores and Paulets. But a still finer specimen of English home architecture is Athelhampton Hall, for four centuries the dwelling of the Martins, whose family derives from Martin of Tours, a companion of the Conqueror. It is a low grey-stone medley of gables, dormers, and high-peaked roofs, mostly belonging to the fifteenth century, with one wing added in the time of Henry VII. There is much good woodwork and glass, and the modern garden-terraces are beautiful. Another lovely manor-house is Bingham's Melcombe, held by the Binghams for about six centuries, apparently built in the time of Edward I. The charm of the building is its picturesque irregularity; it is an ideal homestead, with no disturbing intrusion of modernity. The buttressed gatehouse, still inhabited, is probably as early as Edward II.

At the manor-house of Trent we find the chambers in which Charles II lay concealed for sixteen days after the battle of Worcester, before he escaped by sea. Melbury House, of Melbury Sampford, was erected in the fifteenth century, but was almost entirely rebuilt in 1547 and much altered in modern times. The house of Wimborne St Giles, though dating from the sixteenth century, is perhaps not specially attractive; it is a low quadrangular building, the seat of the Earls of Shaftesbury, and may claim distinction for having at one time sheltered the philosopher Locke. At Winterborne Herringstone is the much-changed home of the Herring family, built in the time of James I. Hanford House, built in 1623, has a fine park of 200 acres on the banks of the Stour; but the old manor of Cranborne claims more attention for its architectural beauty. Probably dating from Henry VIII, the present building owes many of its Jacobean features to the Cecils, and still belongs to the Marquis of Salisbury. Its great hall was formerly used for the periodic courts of the Hundred, the barony, and the jurisdiction of Cranborne Chase. This home was much visited by kings, especially by the Stuarts. At Wool is the Jacobean manor-house of the Turbervilles, now a farm, familiar to the readers of Thomas Hardy. Sandford Orcas has a most attractive Tudor mansion, but the house at Puncknowle has been rebuilt.

Radipole Manor House

Quite a long list of these "haunts of ancient peace" might be given. There are interesting old manor-homes at Up Cerne, Radipole, Chantmarle, Wynford Eagle, Wraxall, Minterne Magna, Wasterton, Poxwell, Tyneham, Rampisham, Toller Fratrum. Many of the farms throughout this charming county were once houses of distinction, and show some surviving feature that speaks of their past. There is one at Godlingstone that dates from the fourteenth century. Perhaps a word should have been given to Hammoon Manor; but it is impossible even to mention all the places that deserve notice. Almost every village can boast of at least one beautiful and dignified old house, one "grey old grange," one ideal English home.

Of modern buildings perhaps the most notable is Bryanston Park, the seat of Lord Portman, near Blandford; it is of red brick and Portland stone. The house at Kingston Lacy is seventeenth century; that of Charborough was built in 1720. The fine Canford Manor belongs almost entirely to the last century, but retains some much older fragments; it is the seat of Lord Wimborne. Modern also are the houses of More Crichel, Milton Abbey, Wotton Fitzpaine, Lytchett Heath, Brownsea Castle, and some others that may seem to claim attention. They are generally on the sites of earlier buildings, relics of which survive in many cases. It is impossible here to deal exhaustively with so wide and interesting a subject.

20. Communications—Past and Present.

Wherever there are settlements of men there will be paths leading from one to another. Even the lower animals, wild and tame, have their accustomed paths and soon wear a track. The earliest road therefore was simply a trodden way, a track worn across moorland and through forest by the feet of men and animals, and later by the wheels of vehicles. Natural instinct would do what it could to assist the passage ; stones would be thrust to one side or perhaps laid down as a rugged paving, impeding branches would be broken away, some attempt would be made to mend a miry or difficult spot. Thus man by slow degrees and hard experience acquired the

Roman Road (from Dorchester to Compton Valence)

art of road-making. Such primitive roads or trackways existed among the Celts, and doubtless earlier ; but road-making was brought to its perfection by the Romans, who made it a branch of their skill in the government of nations. In the early days of tribal strife the inhabitants were more zealous in fortifying their homesteads than in facilitating communication between themselves and others ; but the Roman road had its set military and political

purpose. The Romans designed important main roads
or utilised old tracks between their stations, which had
usually been native strongholds ; thus easing the passage
of the legions, so that no station was isolated but each
in touch with another. By this means imperial domi-
nance was maintained ; the roads, then as now, were
the arteries of social life, but they are less so now, because
their use has been affected by the railways. We all know
at least the names of the great Roman roads—Watling
Street, Ermine Street, Akeman Street ; the Fosse Way
and the Icknield Way. Many of these were founded,
at least partially, on older tracks ; the *Via Iceniana*, or
Icknield Way, in particular, zigzags rather too much
for a true Roman road. It is usually said that it is the
Icknield Way that enters Dorset from Old Sarum (Salisbury), but this is hardly correct. More strictly, it is a
branch from that central highway. The road that entered
Dorset bore different names in different parts of its
course. It led from London to Silchester in Hampshire
where it forked, one branch leading to Winchester and
another to Old Sarum. From Sarum it soon reached the
land of the Durotriges, entering at Woodyates, in the
north-eastern and most vulnerable corner of the county.
Bokerly Dyke, a great rampart of earth running for
miles, still in part the county's boundary, was clearly
thrown up as an obstruction of this very road, which it
once blocked. It has been proved that the road is earlier
than the dyke, and therefore earlier than the Roman
occupation. Crossing the northern Downs to Badbury
Rings, the road became known as the Ackling Dyke ;

from Badbury it proceeded directly to Dorchester, the Roman *Durnovaria*. Traces of it may still be seen at Bagwood and near Tolpiddle. Close to Dorchester, and beyond, it merged in what is now the high road to Bridport. North-west of Badbury there appears to have been another road leading to Hod Hill ; and a continuation

On the Old Coaching Road, Wimborne
"Coach and Horses"

southward led to Poole Harbour. From Dorchester to Ilchester in Somerset there was another important highway, partially surviving in the present road ; and the existing road to Weymouth was yet another branch. From Bridport the road forked to Axminster and Exeter.

The present roads through Dorset are for the most part very satisfactory, good material being plentiful ; but by-roads in the low-lying districts are apt to be miry. One of the principal highways is that from Salisbury to Blandford, Dorchester, and Weymouth, which is direct and well-laid. From Blandford roads branch to Sherborne, Shaftesbury, Wimborne, and Poole; and from Dorchester there are good roads to Wareham, Sherborne, Crewkerne, Portland, and Bridport. From all these there are cross roads too numerous to mention, all clearly shown on the excellent maps that are available. As a rule the cyclist fares well in Dorset, but he must not be afraid of hills, and he must especially be prepared to meet difficulties in the neighbourhood of Lyme and Charmouth.

There are no canals in Dorset, and very little river-navigation exists.

Not being a county of large towns, the railway communications are not so good as might be desired. The north of the county is just touched by the main-line of the London and South-Western from London to Exeter; with which there is a junction at Templecombe, leading by the Somerset and Dorset Railway, now the joint property of the Midland and South-Western, to Blandford, Poole, and Bournemouth. Dorchester and Weymouth are served by the South-Western, and express trains reach Weymouth from Waterloo in a little over three hours. There is a branch at Wareham to Swanage. Maiden Newton, Dorchester, and Weymouth are also reached by the Great Western from Yeovil, with branch-lines to Bridport and Abbotsbury. Lyme is served by a branch from the London and South-Western at Axminster. No fault can be found with the connection

West Bay, Bridport

between London and the chief Dorset centres, but the connection with the north-west, midlands, and Wales sadly needs improvement, and might be greatly helped by a better communication with Bristol. In connection with the Great Western there is a regular sailing of steamers from Weymouth to the Channel Islands, and during the summer months excursion-steamers call at Weymouth, West Bay, and Lyme. There is also a regular steamer service between Swanage and Bourne-mouth.

21. Administration and Divisions— Ancient and Modern.

Of the government and administration of Dorset in the earliest times we can say very little, for very little is known. Druidism, which was largely a legislative system under religious sanction, is supposed by Professor Rhys to be pre-Celtic in origin; and the early government of the Celts, apart from priestcraft, was entirely on tribal lines, with systems of law that were rather elaborate. At the time of the Roman invasion the Britons appear to have been passing from the rule of kings to that of popular chieftains or military leaders, and the separate tribal states (of which Dorset, the land of the Durotriges, was one) were more like despotisms than either mon-archies or republics. When Rome divided our land into Upper and Lower Britain, Dorset fell into the secondary division of Upper Britain known as Britannia Prima, with a military centre of control at Dorchester (*Durno-varia*). Roman Britain was a province, or perhaps rather a diocese of five provinces, under a vice-prefect or vicar, who was responsible to the praetorian prefect of the Gauls. Three chief generals divided the military command. After the departure of the Romans there must have been a partial lapse to former conditions; and it cannot have

St Peter's Church and Town Hall, Shaftesbury

been before 700 A.D. that the West Saxons really gained absolute control of the district that lay south-west of the great Selwood forest. By this time the Saxons had been Christianised, and brought a somewhat milder rule. The division of south England into shires, attributed to Alfred, made Dorset a shire (a share or shearing) of Wessex, with its own ealdorman or earl, and its shire-reeve or sheriff, who was the direct representative of the

king for purposes of taxation. In 705 Sherborne became
the see-town of a bishopric extending from the English
to the Bristol Channel. The shire was subdivided in
Hundreds, with a monthly court or Hundred-gemot;
there was also a Folk-Moot, usually half yearly, presided
over by ealdorman, sheriff, and bishop. Dorset consisted
of 33 Hundreds and 21 Liberties, "liberty" being an old
term signifying districts of exemption from the jurisdiction
of the sheriff. These divisions still survive, though not
the privileges of the Liberty. Besides these there were
the townships, each with its town-reeve, some being free
towns, others dependent on a feudal lord. A higher
class of township was the "burh," the residence of
king, feudal lord, or chief magistrate (who later became
the mayor). When the middle classes began to rise in
importance, there were trade-guilds and craft-guilds,
sometimes growing really powerful. The villages and
hamlets were usually held, in feudal manner, by an
esquire, whose title still survives when we give the term
squire to a country landlord.

Bridport has an unusually well-preserved collection
of ancient municipal deeds dating from the thirteenth
century.

Till quite recent years Dorset sent fourteen members
to Parliament; it now sends only four, being divided into
four parliamentary sections, north, south, east, and west.
Its magistracy is presided over by a Lord Lieutenant and
Custos Rotulorum, with deputies and local magistrates.
There are Poor Law Unions at twelve different centres.
The Police, under control of a committee of Quarter

Sessions and County Council, has its head-quarters at
Dorchester ; it consists of one chief constable, one deputy,
10 superintendents, 24 sergeants, and 154 constables.
The county falls under the Western Circuit, and has
courts of Quarter Session at Dorchester and Poole, with
Petty Sessional courts at Blandford, Bridport, Cerne,
Dorchester, Shaftesbury, Sherborne, Wareham, Stur-

Town Hall, Poole

minster Newton, and Wimborne.   Poole is a county in
itself.   There are 285 civil parishes.   Ecclesiastically, the
county falls into the Salisbury diocese, with Rural
Deaneries of Dorchester, Bridport, Shaftesbury, Pimperne,
and Whitchurch. Blandford, Bridport, Dorchester, Lyme,
Poole, Shaftesbury, Weymouth and Melcombe Regis,
and Wareham, are municipal boroughs.   The general

administrative business of the county is now transacted by
a County Council, consisting of a Chairman, 19 Aldermen,
and 57 Councillors ; and there is a separate County
Council Education Committee, with 34 members.

22. The Roll of Honour of the County.

Dorset is a county of such small population—the
entire population is only that of a fair-sized town—that
we could not be surprised if its list of worthies proved a
very short one. This is not the case ; the list is very
fairly full and contains many notable names. It seems to
be a fact that a given number of people scattered abroad
over a county will produce more persons of distinction
than the same number crowded together in a city ; the
scattering, the greater isolation, give freer play to character
and individuality. We come across more persons, pro-
portionately, that may be described as "characters" in
country places than we do in cities. Dorset has its full
share, and always has had.

Beginning with ecclesiastical personages, we may claim
the great West Saxon Aldhelm, or Ealdhelm, as most
intimately associated with Dorset, though the place of
his birth is unknown. After building churches at Frome
and at Bradford-on-Avon—the latter of which is claimed
by some as identical with the present building, though
other authorities will not admit an earlier date than 900—
he had a large share in the founding of Malmesbury
Abbey, and became first Bishop of Sherborne in 705.

He was of royal blood, and a man of great learning and piety. St Aldhelm's Head, south of Swanage (sometimes erroneously named St Alban's Head), is named after this great prelate. The "Grammarian" Aelfric, who translated Latin homilies into excellent Anglo-Saxon for the use of his monks, was the first abbot of Cerne ; and Cardinal Morton, who was raised to the

Bottom Valley and St Aldhelm's Head

Archbishopric of Canterbury by Henry VII, had also been a monk at Cerne. Another primate, the distinguished William Wake, was born at Blandford in 1657 ; and contemporary with him was Thomas Lindesay, Archbishop of Armagh, who was also born at Blandford. Frampton, Bishop of Gloucester, one of the "Seven Bishops," was born at Pimperne. Thomas Sprat, Bishop of Rochester, whose memoir we find in Johnson's *Lives*

of the Poets, was born at Beaminster in 1635, and John
Hutchins, the native historian of Dorset, was rector of
Wareham.   The Catholic Cardinal Weld belonged to a
Lulworth family, and the eminent Nonconformist John

William Barnes

Angell James was a native of Blandford.   Stillingfleet
was born near Cranborne.

Leaving this long list of divines we may pass to
William Barnes, who claims distinction as a poet rather
than as a clergyman.   He was born at a farm near

Pentridge, close to the Wiltshire border, in 1801, but passed his boyish years at Bagber and Sturminster Newton. After entering a solicitor's office he became a schoolmaster. He was ordained in 1847, and spent the last twenty-five years of his life as rector of Winterborne Came, near Dorchester. After writing much on philological and archaeological subjects, he published his first poems in Dorset dialect in 1844. He died in 1886. Dialect was a perfect medium for his genius, limited in its range but genuine in quality; and his poetry is distinguished by its freshness, humour, and pathos, and a power of depicting nature so artless as to pass all art. Apart from Barnes, Dorset cannot boast highly of the quality of its native-born poetry. Wimborne lays claim to Matthew Prior, but somewhat doubtfully. But if the poets have not been born in Dorset, they have visited it and loved it. Wordsworth with his sister spent two years at Racedown, on the slopes of Pilsdon, and was visited here by Coleridge. Thomson and Young were visitors at Eastbury Park, whither also came Voltaire, Bentley, and other notable persons. Blandford gave birth to two useful translators of classic poetry, Christopher Pitt and Thomas Creech, whose work has been long superseded. Pope was a guest at Sherborne, whose castle was at one time held by Sir Walter Ralegh; and the poet Wyatt lies buried in its Abbey.

Among other noted residents must be mentioned the novelist Fielding, who inherited the manor of East Stower from his mother, and wasted his wife's small fortune by extravagant living. His immortal " Parson Adams " was drawn from the Rev. W. Young, Vicar of West Stower. Robert Boyle, physicist, founder of the "Boyle Lectures," and fourteenth child of the 1st Earl of Cork, resided and made his first scientific experiments at Stalbridge Park, which he inherited from his father; Jeremy Bentham lived at Ford Abbey, and was visited there by the Mills. John Locke was a tutor of the philosophic Lord Shaftesbury at Wimborne St Giles, which had also been the birthplace of the famous Anthony Ashley Cooper, 1st Earl of Shaftesbury, member of the "Cabal" ministry. The great William Pitt spent happy boyish days at Weymouth and at Lyme, and Lord Chancellor Eldon ended his days at Encombe, his family seat on the Isle of Purbeck.

Lyme Regis can boast the birth of Sir George Somers, coloniser of the Bermudas or "Somers' Islands"; and many will think with pleasure of the association of this town with Jane Austen, who was remarkably fond of the place, and introduces it in her novel *Persuasion*. Another interesting literary association is that of Weymouth with Fanny Burney; and it is well also to remember that this popular watering-place can number Christopher Wren and Joseph Hume among its members of Parliament, in the days before it was disfranchised. Admiral Sir Thomas Hardy, one of Nelson's ablest colleagues, was born at Portisham, and a column to his memory has been erected on Black Down. Admiral Viscount Hood was a native of Thorncombe, his father (another of whose sons was the first Lord Bridport) being master of Beaminster School. Captain Coram, born at Lyme about 1667, though a sailor, is less distinguished as such than as a philanthropist, for it is to him we owe the Foundling Hospital, which he lived to see opened in his old age, in 1741. Another famous naval character was Harry Page or "Arripay" of Poole ; and this town also gave birth to the eminent naturalist, Thomas Bell.

Lastly we may claim Alfred Stevens, one of the greatest of English sculptors, as a Dorset man. He was born at Blandford the son of a house-painter in 1818 and died in 1875.

Before quitting this subject, some mention should be made of the part Dorset played in the foundation and success of New England. John White, rector of Dorchester, had much to do in furthering and encouraging the sailing of the Puritans ; and in 1628 John Endicott and other Dorchester gentlemen sailed from Weymouth in the "Abigail," the result of which expedition was the establishment of a new Dorchester across the Atlantic. It is said that the Channings also came from the English Dorchester, and the Washingtons from Cerne Abbas, thus giving Dorset a prominent place as the old home of some of America's most noteworthy men.

Besides these, the novelist and poet Thomas Love Peacock was born at Weymouth ; the poet Crabbe was at one time vicar of Evershot ; and Motley, the American historian (born in Dorchester, Mass.), resided at Dorchester and died at Kingston Russell.

23. THE CHIEF TOWNS AND VILLAGES
IN DORSET.

(The bracketed figures after each name give the population in
1901; the figures at the end of each section give references
to pages in the text. Architectural abbreviations are *Perp.*,
Perpendicular; *Dec.*, Decorated; *E.E.*, Early English; *Norm.*,
Norman.)

Abbotsbury (707) is situated at the junction of the Chesil
Bank with the mainland, about eleven miles from Portland. There
was a Saxon Abbey here, of which some remains survive; the
most striking of which is the great tithe-barn, which is 276 feet
in length, and has the appearance of a church. The castle is
modern, but Abbotsbury Camp, about one-and-a-half miles
distant, is a fine earthwork, and there is also a charming gabled
manor-house. At the Dissolution the Abbey passed to the
Strangways, whose house, consisting of the abbey buildings, was
destroyed by the Parliamentarians in the Civil War. The church,
chiefly Perp., has a curious plaster ceiling displaying the Strangways' arms. St Catherine's Chapel dates from the fifteenth century.
Below extends the West Fleet with its famous Swannery and Decoy.
The swannery, which belonged to the abbey, is the home of over
1000 swans and numerous wild-fowl. (pp. 38, 41, 96.)

Affpiddle (358) is a typically beautiful Dorset village,
eight miles north-east of Dorchester, whose church (Perp. with
E.E. chancel and Norm. font) has fine woodwork of the sixteenth
century, as recorded by a quaint inscription on one of the seats.

St Aldhelm's Head, corruptly named St Alban's, is 440 feet in height, with an old chapel dedicated to St Aldhelm, the great Wessex bishop. Like St Catherine's, this was a sailor's oratory. The doorway is Norman. (pp. 10, 33, 43, 50, 121.)

Athelhampton (62), about six miles north-east from Dorchester, contains one of the finest old Dorset manor-houses, Athelhampton Hall, for four centuries the home of the Martins.

Tithe-barn, Abbotsbury

The "new" wing was built in the time of Henry VII, to which date belong its panelled state-bedroom and the exquisite banqueting hall, with trefoiled oak roof and oriel window. Some earlier parts date back to the time of Edward III. A fine gatehouse and chapel were destroyed about 50 years since. Athelhampton is supposed to take its name from a former palace of Athelstan's here—which is possible but doubtful. (p. 107.)

Batcombe is a tiny hamlet in the solitude of the Downs, three miles north-west of Cerne Abbas, with a lonely low-lying church associated with memories of a former squire, Minterne the Conjuror. On Batcombe Downs is a curious monolith named the Cross-in-hand.

Beaminster (1702) is a town whose many fires have marred its genuine antiquity, causing it to be more than once rebuilt.

Beaminster from the South-West

The beautiful pinnacled tower of its church is late Perp., richly decorated; there are some E.E. features in nave and south-aisle. The fine manor-houses of Parnham, Mapperton, and Melplash are near. The town is now an active centre of agriculture, and famous for the production of "Blue Vinny" cheese. (pp. 69, 98, 106, 122, 124.)

Bere Regis (1404), a large village on the borders of the Great Heath, was once a royal residence, with a manor belonging to Elfrida, the murderess of Edward the Martyr. King John also stayed here during some of his numerous visits to Dorset; later the manor belonged to Simon de Montfort and to the Turbervilles. The church, with its Turberville tombs and window, is one of the best in the county; its magnificent timber roof, decorated with figures supposed to represent the Apostles, was a gift of Cardinal Morton, who was born here and who also gave the western tower. The Turbervilles were a notable family, but they owe their modern fame to the genius of a living writer. (p. 98.)

Bindon Abbey is the ruin of a Cistercian monastery, in the woods by the side of the Frome, near the road from Wareham to Dorchester; it was founded in 1172. The remains have largely gone to supply building material, and the bells to other parishes. (p. 96.)

Blandford (3550), consisting of Blandford Forum and Blandford St Mary, is an old town whose many fires have rendered it a modern one. The church of Blandford St Mary is Perp., with some earlier remains. Governor Pitt, grandfather of the first William Pitt, was buried here. Ryve's Almshouse, 1682 (the Ryves were a notable local family), survived the many fires. Blandford is now a pleasant, active market-town, watered by the river Stour. Aubrey the antiquarian went to school here, and here were born the translators Pitt and Creech, Wake, Archbishop of Canterbury, and Lindesay, Archbishop of Armagh. Gibbon, who visited here, speaks affectionately of "pleasant hospitable Blandford." (pp. 69, 110, 114, 121-3, 125.)

Bradford Abbas (391), the "Broad ford" over the Yeo, once belonged to Sherborne Abbey. The church is famous for its buttressed and pinnacled tower and its beautiful west front.

Branksome (2812) is a fashionable suburb of Bournemouth, and shares that town's modernity.

S. D.   9

Bridport (5962), the port of the Brit or Bredy, is close to the Roman road from Dorchester to Exeter, and had become a place of importance in Saxon times. Its harbour is now generally called West Bay; it can only accommodate vessels of 250 tons; the coasting and timber trades are no longer what they were. As some recompense, West Bay is becoming a watering-place; though not specially attractive in itself the district is charming. Bridport is an interesting old town, of no great historic memories, unless its share in the escape of Charles II be called so. (pp. 69, 118.)

Broadwindsor (994), three miles north-west of Beaminster, and close to Pilsdon Pen and Lewesdon Hill, is associated with Thomas Fuller, who was rector here; his church, much restored, retains its Norm. and E.E. aisles and Norm. font, together with Fuller's pulpit. To the west of Pilsdon is Racedown Lodge, where Wordsworth and his sister resided for nearly two years (1795-97). His sister writes enthusiastically of the "lovely meadows above the top of the combes, and the scenery on Pilsdon, Lewesdon, and Blackdown Hill, and the view of the sea from Lambert's Castle." It was here that Wordsworth wrote his "Ruined Cottage" (one of the finest portions of the *Excursion*), and also his play "The Borderers"; here also he was visited by Coleridge. (p. 123.)

Burton Bradstock (574), one of the most picturesque villages in Dorset, at the commencement of the Chesil Bank, was a manor presented to St Stephen's, Caen, to redeem the regalia which William I had bequeathed. At one time it belonged to Bradenstock Abbey, Wiltshire, whence its second name; being on the Bride river, the Burton is a corruption of Bridetown.

Came or **Winterborne Came** (103) is chiefly interesting for the fact that William Barnes, the Dorset poet, was rector here from 1862 till his death in 1886. It is a wide, thinly-populated parish, two miles south-east of Dorchester, with a charming little church in whose graveyard the poet lies. (p. 123.)

Canford Magna (1524), on the Stour, six miles north-east of Poole, contains Canford Manor, the seat of Lord Wimborne; the house was built about 80 years since. (p. 110.)

Cerne Abbas (643) has some remains of its Benedictine Abbey, founded in 987 (and even then on the site of an older foundation), of which Aelfric the Grammarian was the first abbot. The abbey was pillaged, but afterwards enriched by Canute. The beautiful gatehouse with two-storied oriel window survives. Cerne Church is Perp., with an imposing tower. Above the town is the Cerne Giant, a figure cut in the chalk after the manner of the different White Horses. The giant is 180 feet in height, and is supposed to be of prehistoric date. (pp. 79, 85, 96, 121, 125.)

Charminster (1679), two miles north-west of Dorchester, contains the County Lunatic Asylum, which accounts for about half of its population. It is also notable for Wolfeton House. (p. 104.)

Charmouth (560) has been named as a Roman station, and was certainly close to a Roman road. Its chief historic memories however are of troubles with the Danes, who chose it as a convenient point of attack, burning and pillaging rather than settling; they left no traces though we know that they utilised the British Coney Castle as a stronghold. In a literary sense Charmouth, like Lyme, has memories of Jane Austen, who visited here in 1804. The district is hilly and very beautiful. (pp. 77, 81, 87.)

Chideock (551), beautifully situated close to the Devon border, some two miles west of Bridport, has some slight trace of the old castle of the De Chideocks destroyed in the Civil War. There is a richly decorated Catholic chapel in the grounds of the manor.

Corfe (1102) was anciently named *Corvesgate*, from the Anglo-Saxon *ceorfan*, to cut or carve; it is a picturesque village in a gap in the Purbeck Heights dominated by the remains of its

Gate-house, Cerne Abbey

fine castle, one of the noblest Norman ruins in England. The castle is famous as the scene of the murder of the Saxon king Edward by his stepmother Elfrith or Elfrida. The king had been hunting, and called at the castle, then a hunting-lodge, for refreshment, and tradition says that he was stabbed by Elfrida herself, while drinking the wine that she had brought him. He was buried first at Wareham and then at Shaftesbury, and was succeeded by his stepbrother Ethelred, for whose sake the murder

Charmouth Beach

was committed. The castle experienced stirring times during the Civil War. Corfe village remains unspoiled and delightful, with its attractive restored church, graceful market-cross, and quaint dwelling-houses. (pp. 78, 100.)

Cranborne (687) is named from the Crane-burn, and was formerly of such importance as to give its name to the great tract of forest, still partly surviving, known as Cranborne Chase, a part of the larger Selwood. Much of the Chase is in Wiltshire, and it formerly extended to Hampshire also, with a circuit of about 100 miles. The shire-boundary and the Old Sarum road both ran through it. It became a great haunt of smugglers, poachers, and vagrants of all kinds, and was disafforested in 1830; before which date it was estimated to contain more than 12,000 deer. The quiet little town has two great glories, its noble church, surviving from a Benedictine priory, and its manor-house. (p. 12.)

More Crichel (362), six miles south-west of Cranborne, contains Crichel House, the seat of Lord Alington.

Dorchester (9458) may claim to be one of the oldest towns in England; one of its streets, the Icen Way or road of the Iceni, recalls Celtic times, while another, Durngate, reminds us of the Roman name for the city, *Durnovaria*, the Saxon Dornceastre. The "Walks" mark the position of the Roman walls, while the near entrenchments, Maiden Castle, Poundbury, etc., speak of prehistoric occupation. The amphitheatre, from recent excavations, seems to have been a site in still more remote times. The town derived its name, as did the county, from the *dwr* or water (the river Frome), and there are signs that it was occupied before the coming of the Durotriges whose name took this same derivation. There is another Dorchester, in Oxfordshire, and the Saxons distinguished the two by terming the Dorset town *villa regalis* and the Oxfordshire town *villa episcopalis*. There are stories of pillaging by the Danes; a Norman castle once stood on the present site of the prison, and a Franciscan friary was near the existing Priory Lane. In 1613 there was a great fire, destroying more than 300 houses. Like some other Dorset towns, Dorchester has a long list of great fires. In the Civil War the town was strongly Parliamentary in sympathies, suffering severely in consequence from the pitiless Prince Maurice. Cromwell was here in 1645. Its Puritans had already supplied many emigrants to New England, where a new Dorchester was

Roman Amphitheatre, Dorchester

founded. At the "Bloody Assize" after the Monmouth rebellion more than 300 persons were tried, and 292 received sentence of death, commuted in many cases to slavery in the Plantations. Of the churches, St Peter's has a fine Perp. tower, and is supposed to stand on the site of a Roman temple. A bronze statue of the poet Barnes stands without, close to the County Museum. The Grammar School was founded in 1569 by Thomas Hardy, ancestor of Admiral Hardy and of the novelist. The town is now much modernised. Fordington is now a suburb. It has a very fine church tower. (pp. 4, 7, 55, 86, 87, 113, 117, 125.)

Ford Abbey historically belongs to Devon, but was transferred to Dorset with the parish of Thorncombe. The Cistercian house was founded in 1141. The third abbot, Baldwin of Exeter, became Archbishop of Canterbury; another abbot, John of Devon, confessor of King John, was a learned theologian; and Charde, last of the abbots, was a great builder. Much of the existing buildings was due to his labour, his Tudor Perp. being remarkably beautiful. Cloister and Abbots' hall remain unspoiled. Some new work was done by Inigo Jones in the seventeenth century. Various holders have had the abbey since the Dissolution, and about a century since it was tenanted by Jeremy Bentham. The house contains the exquisite Raphael tapestries, worked from the original cartoons, for which Catherine of Russia is said to have offered £30,000. The beautiful present chapel was the monastic chapter-house. (p. 98.)

Frampton (378), its name a corruption of Frome-town, is a pleasant riverside village about five miles north-west from Dorchester. The manor was connected with the Sheridan family, and has memories of Lady Dufferin and Lady Caroline Norton. The house dates from 1704. British and Roman remains have been discovered here. (p. 21.)

Gillingham (3380), on the Stour, four miles north-west of Shaftesbury, once comprising a royal park, forest, and hunting-lodge, is now a modernised active little town of breweries, potteries, sacking-works, and bacon-curing. The Saxon kings came here, and in 1042 a witan was held. At Slaughter Gate, just by, Canute was defeated by Edmund in 1016. Gildon, who figures in Pope's *Dunciad*, was born here, and the historian Clarendon was a scholar at the grammar-school, founded in 1516. (p. 69.)

Hilton (502) is a secluded little place seven miles south-west of Blandford surrounded by heights of the down-land—Bulbarrow, Nettlecomb Tout, and others, with their prehistoric earthworks. There is a curious panel-painting of the Apostles in its church, taken from Milton Abbey. (p. 84.)

Horton (331), four-and-a-half miles south-west of Cranborne, was formerly a Benedictine Abbey; its church contains an effigy of Giles de Braose, who died in 1305, and of his wife. At "Monmouth's close," a field a few miles distant, the unhappy Duke of Monmouth was captured after his defeat at Sedgemoor. (p. 81.)

Iwerne Minster (543) is the seat of Lord Wolverton; its church is notable for one of the few stone steeples to be found in Dorset—there are only three in all. This and Iwerne Courtenay are named from the little river Iwerne. (p. 98.)

Kimmeridge (338) is chiefly noteworthy as giving its name to the oolitic formation known as Kimmeridge Clay. (See Geology.) There was an attempt to work this into gas and oil, and a small quay was constructed on the coast, but the effort failed. (pp. 32, 74.)

Kingston (399), near Corfe, has two churches, the older rebuilt by the first Lord Eldon, who lies buried within it, and the second built by the present Lord Eldon. Near is Encombe.

Lulworth, East (294), named *Lulworde* in Domesday and Lilleworth by Leland, lies just west of the Isle of Purbeck. There are two castles, the British "Flowers Barrow" and that completed by the Welds in the seventeenth century. Cardinal Weld, the founder of the Jesuit college at Stonyhurst, was of this family. The French King, Charles X, was a resident here in 1830.

Lulworth, West (426), is chiefly noteworthy for the beautiful little Lulworth Cove, a favourite haunt of sight-seers. (pp. 44, 74, 84, 104.)

Lyme Regis (2095), developing from a quiet fishing-village to a popular and most beautiful watering-place, takes its name from the river Lyme. It has been identified with the Roman *Londinis*, and the coast-road from Dorchester to Exeter passed here. In 774 a grant of the land was made to Sherborne by Cynewulf. The town was enfranchised by Edward I; it supplied four vessels to the siege of Calais, and in return suffered much from French attack. Two vessels from Lyme sailed to meet the Armada, the first onset of which was witnessed from the heights. The town stood a severe siege in the Civil War, losing only 120 men while the Royalists lost 2000. Parliament rewarded Lyme with a grant of £2000. Cosmo de Medici landed here in 1669; and in 1672 a fight took place between English and Dutch vessels off the coast. On June 11, 1685, the Duke of Monmouth landed at Lyme and was warmly welcomed by the people, who were strongly Protestant and largely Puritan in their sympathies. Among his recruits was the novelist Defoe. This ill-fated rising, beginning thus in Dorset, ended in Dorset by the capture of the Duke near Horton, and the county had to pay heavily for its enthusiasm. Thirteen victims were executed at Lyme. The town has had some notable residents and visitors. Sir George Somers was born here in 1554 and Thomas Coram, the founder of the Foundling Hospital, more than a century later; also Case, a noted quack and astrologer of the time of James II. Young William Pitt spent some months here in 1773, and in the early nineteenth century the town won the affection of Jane Austen,

Lyme Regis, Sherborne Lane

who has immortalised its Cobb in her *Persuasion*.   Miss Mitford
too resided here in her youth.   The church is specially interesting;
it is chiefly Perp., with an earlier tower and some Norm. work.
Traces of the old municipality survive in the Court Leet and
Court of Hustings.   The Cobb, a semicircular stone pier, probably
dating from Edward I, has been often destroyed and often rebuilt,
its last reconstruction being in 1825.   The Lyme coast is noted
for its fossils and its landslips.   It was here on the Charmouth

Milton Abbey

side, that the girl Mary Anning discovered the famous remains
of the ichthyosaurus now in the South Kensington Museum.
(pp. 47, 69, 73, 79, 81, 114, 124.)

Marnhull (1286), pleasantly situated in the Vale of Blackmore, has a good church, containing features of all periods from
Norm. to Perp.   Its alabaster effigies, somewhat mutilated, are
supposed to be of Thomas Howard and his two wives.   Nash
Court, near, is Elizabethan.   (p. 98.)

Melcombe Bingham (136), near Milton Abbey, has a most charming manor-house. (p. 108.)

Milton Abbas (650) is curious for the fact that in the middle of the eighteenth century the village was entirely transplanted to its present site by its landlord, the first Earl of Dorchester, who wished greater privacy for his mansion Milton Abbey. The cottages, placed in pairs, with fine chestnut trees between each, form a beautiful and striking avenue. The abbey church is

The White Horse, Osmington

almost worthy to rank as a cathedral for its beauty.   St Catherine's Church, in the abbey grounds, dates from 938.   It was formerly desecrated, but has been restored and is again used. (p. 96.)

Osmington (334) four miles north-east of Weymouth has a gigantic figure of George III on horseback cut in the turf. Osmington Mills, on the sea coast, has become popular with visitors.   (p. 149.)

Piddletown (934), which is some five miles north-east of Dorchester, has some quaint and beautiful houses, and an unspoiled church with panelled chestnut roof. The singing gallery bears the date 1635, and there are six ancient effigies, one dated 1250 and a very fine one of alabaster 1475. The tower is late Norm. with additions.

Poole (19,463) has been claimed, rather dubiously, as a Roman station. It certainly shared Wareham's exposure to Danish pillage. Incorporated in 1248, the town was made a distinct county by Elizabeth. The inhabitants showed much of the bold privateering spirit so notable in Channel coast-towns of the past, being generally engaged in marauding expeditions to the Continental shores, or in a combination of piracy and smuggling. One specially dauntless and successful pirate was the famous Harry Page, known to the French as "Arripay," who is said to have once returned with as many as 120 prizes from the Breton coast. He so worried the opposite shores that in 1406 a joint expedition was despatched to Poole against him by the French and Spanish kings; but though driven to refuge on the Great Heath, Arripay was not taken. The town was strongly Parliamentarian in the Civil War, and successfully resisted attack. There are some relics of old maritime Poole around the quays. (pp. 18, 22, 38, 41, 47, 50, 69, 72, 119, 125.)

Portisham (582), near Abbotsbury, has a lofty monument to Sir Thomas Hardy, Nelson's colleague at Trafalgar. (p. 124.)

Portland (15,199) is now wrongly named an island, but its outlying mass of oolitic limestone, having resisted the action of the waves that have denuded the neighbouring coasts, was once really insular, and the Chesil Bank that unites it to the mainland is geologically of quite recent formation. This barren headland, known to the Romans as *Vindilis*, is now little more than a vast quarry, with some dusty settlements and fortifications. One of the earliest Danish descents took place here, but in times when defence was better understood Portland was fairly well able to hold its own. The castle was built by Henry VIII; it was taken for the king in the Civil War, but surrendered in 1646. A naval battle was fought off the island in 1653 between the Dutch under Tromp and De Ruyter, and the English under Blake and Monk. The battle was stubbornly contested, but the English remained victors though Blake was seriously wounded. Portland is a royal manor and a "liberty," with its own Reeve and some peculiar local customs. Some of the quarries belong to the Crown, others are leasehold; and a large portion of the isle has been transported bodily, in the shape of Portland stone, to London and elsewhere, the value of this building material being first appreciated by Inigo Jones and Wren. The great modern breakwater, taking 23 years to construct, was chiefly the result of convict labour. About 700 convicts are usually in residence here. Most of the stone villages (Fortune's Well, Castleton, Easton, Weston, Chesilton) are frankly ugly. The chief fort is at Verne. There is a fine new lighthouse on Portland Bill. (pp. 31, 45, 48, 50, 67, 70, 71, 73, 79, 87, 103.)

Preston (664), three miles north-east of Weymouth, has an early Norman bridge, once considered Roman.

Puncknowle (363), midway between Bridport and Abbotsbury, has a beautiful little manor-house near the church. (p. 109.)

Shaftesbury (2027), locally called Shaston, is just within the borders of the county. It is a most ancient town, on the site of a prehistoric settlement—the Mount Paladur of Geoffrey of Monmouth. The Romans occupied the earthwork at Castle Hill; and about 888 Alfred founded the abbey for Benedictine nuns, his daughter Ethelgiva being its first abbess. The Wessex kings richly endowed this abbey, and after Edward the Martyr was buried here it became a great place of pilgrimage owing to the cures wrought at his tomb. Canute died here in 1035, but was buried at Winchester. At the Dissolution the estate passed to the

Earl of Southampton. From its few remains it is still possible to
guess at the wealth and magnificence of this abbey. Besides this
flourishing establishment Shaftesbury is reputed to have had
twelve churches, three mints, and two hospitals, besides many
chantries; but the only complete survivor is St Peter's Church.
Shaftesbury is now a small, quiet country town standing on a
ridge of the chalk downs, at a height of about 700 feet. (pp. 78,
79, 85, 96.)

Gold Hill and Part of Old Walls, Shaftesbury

Sherborne (5760) is supposed to derive its name from the
Anglo-Saxon *scir-burn*, the "clear brook"; but it is spelt Schire-
burn in King Alfred's will, and as the Yeo is a kind of natural
boundary of the county, *shire*-burn may be its true meaning. One
of the first acts of the West Saxons on gaining supremacy west of
Selwood, was to found a bishopric, with its see-town at Sherborne,
in 705. The early history of the town is wrapped up in that of
the abbey, which had some notable bishops, including Aldhelm

and Asser. The old Sherborne bishops were men of action as well as of learning and piety; they fought bravely, and three of them fell in battle against the Danes, who never could make much lasting impression on this corner of Wessex, though they harassed it unmercifully. The present abbey is on the site of Aldhelm's building, of which it may enshrine a few possible remains. The school, said to have been founded at the same time as the abbey,

The Dining Hall, Sherborne School

did not suffer at the Dissolution, and was fostered by Edward VI, whose statue is in the dining-hall. There is a fine old almshouse, refounded in 1437. The groined Abbey Conduit was built by Abbot Frith in the fourteenth century. The old and modern castles have been spoken of elsewhere. Stephen Harding, one of the founders of the Cistercian order, was born and educated at Sherborne. (pp. 79, 94, 102, 123.)

Stalbridge (1504), formerly Staplebridge, is a centre of the Blackmore Vale dairy district; it has a striking fourteenth century stone cross and a restored ancient church. Many Roman coins were found in a gravel-pit here. (p. 124.)

East Stower or **Stour**, four miles west of Shaftesbury, is noteworthy as having been the residence of the novelist Fielding, who lived here with his first wife and quickly ran through the fortune she had brought him; the house which he inherited from his mother has been rebuilt. (p. 123.)

Studland (474) is a delightful little coast-village on a beautiful bay, too attractive to remain long unspoiled. The church is a charming example of good Norman work. On Studland Common is the Aggleston or *Heligstan* (Holy Stone), a block of iron-ored sandstone, weighing 400 tons, which tradition says that the devil threw across from Portland. Many have wondered how it was brought here; probably it was not brought at all, being simply a mass of rock exposed by the weathering and denudation of softer measures. None the less it was doubtless venerated as a holy stone in Saxon times and earlier. (p. 43.)

Sturminster Marshall (721), apparently the elder of the two Sturminsters, is mentioned in the will of King Alfred. But the same connection is claimed for the other "minster on the Stour."

Sturminster Newton (1877) was much favoured by Saxon kings, who had a castle or hunting-lodge here, now vanished. The manor was given to Glastonbury Abbey. There is an attractive old bridge connecting Sturminster Marshall with the "new town." William Barnes was educated at the school here. (pp. 104, 123.)

Swanage (3408), the Saxon Swanwich, or Swanwic, shows a considerable growth of population since the 1891 census. It can boast of a naval victory of Alfred over the Danes in 877, the first

The Globe, Durlston Head, Swanage

English sea-victory on record. Old Swanage, which carved its
fortune out of its stone, has been very much modernised; even
the church is modern with the exception of its rugged tower.
The town has both gained and lost by the nearness of its Hampshire neighbour Bournemouth; it gains financially by the visits
of thousands of trippers who cross by steamboat, but it loses by
the greater fashion and entertainment that the less beautiful
Bournemouth has to offer. Even those who care little for natural
beauties are amused by the huge terrestrial globe, constructed
from a mass of Portland stone, on Durlston Head. (pp. 38, 43,
68, 74.)

Tarrant Gunville (303) is one of the many Tarrants
taking name from a tributary of the Stour. In this parish is
Eastbury House, formerly belonging to George Bubb Dodington
with whom Browning "parleys" in one of his poems, the friend of
Thomson, Young, Fielding and Voltaire. Tarrant Rushton has
an interesting church dating from 1150. Tarrant Crawford, with
an unspoiled rustic church, was the burial-place of Joan, Queen
of Alexander II of Scotland. Above Tarrant Keynston is a fine
Celtic earthwork, now a golf-course. (p. 123.)

Trent (354), two miles north-east of Yeovil and formerly
in Somerset, is associated with the escape of Charles II, who lay
concealed at the manor-house. The church deserves mention
for its fine bench-ends and rood-screen, effigies, and Dec. tower.
(p. 108.)

Upwey (812) is visited by thousands of Weymouth tourists
for the sake of its Wishing-Well; the village is beautifully
situated and has an interesting church.

Wareham (2003), lying between the rivers Frome and
Trent at the head of the Poole estuary, is an ancient walled
town, its rectangular walls being of earthwork instead of masonry.
It was doubtless held as a settlement and stronghold before the

Romans, probably by the Belgic Morini, whose name survived in the Roman *Morinio* with which Wareham is identified. It was much troubled by the Danes, and was destroyed by Canute; it also suffered severely in the war between Stephen and Matilda, and in the Civil War, when its rector, William Wake, grandfather of Archbishop Wake, was very badly used by the Parliamentarians. Modern Wareham has shrunk within its walls. There was a Norm. castle on Castle Hill. Close to the picturesque, but now almost deserted, quay is the church of St Mary, in which Edward the Martyr was buried before being taken to Shaftesbury. Old as this church is, it contains some inscribed stones that point to an earlier building, possibly Celtic; and a stone with a Norse inscription recalls the Danish occupation. In the St Thomas Becket chapel is a remarkable five-light Roman cresset and an early lead font. Much of Wareham was destroyed by fire in 1762; but the church of St Martin, built on the walls, with a Norm. chancel-arch and a Saxon window still exists. Beohtric, a West Saxon king, is said to lie here. Hutchins, the Dorset historian, was a former rector of St Mary's. (pp. 18, 47, 69, 77, 78, 98.)

Weymouth (19,843) is the name given to the combined boroughs of Weymouth and Melcombe Regis, which at one time were distinct. The two neighbours quarrelled bitterly till the time of Elizabeth, when a bridge was built to connect them and they were joined by act of Parliament. They both returned two Members till the Reform Bill. Jordan Hill is identified with the Roman *Clavinium*, with Radipole for its port. Weymouth was of some importance in Saxon times. It sent 20 ships to the siege of Calais, and six to meet the Armada. The town suffered considerably during the Civil War. Its modern prosperity dates from 1789, when George III visited it for sea-bathing with his queen and princesses. The parts of old Weymouth around the harbour are still interesting, but modern Weymouth is simply a typical watering-place with a bright sea-front and beautiful surroundings. The green headland known as the Nothe has been strongly fortified. (pp. 45, 55, 69, 71-2, 79, 104, 116, 124, 125.)

Whitchurch Canonicorum (671), once belonging to the canons of Wells and Salisbury, the *Witan-cerce* of Alfred's will, is in the Vale of Marshwood. It has a fine church dedicated to St Wita or Candida, one of the most interesting in Dorset, being of the early Gothic styles of Wells and Glastonbury. (p. 100.)

Wimborne Minster

Wimborne (3696), doubtfully identified with the Roman *Vindogladia*, was certainly a Romanised settlement, and became important in Saxon times, when its minster was founded, probably about 705, by Cuthberga, sister of King Ina. Ethelred was buried here, but this first religious foundation was entirely destroyed by the Danes. It was succeeded by a collegiate church about the year 920, and so it continued to the Reformation, its deans being usually pluralistic churchmen of high position.   One of these deans was the young Reginald Pole, afterwards Cardinal. The grammar-school was established by Margaret, mother of Henry VII.   St Margaret's Hospital seems to date from the early thirteenth century.   There is little to be seen at Wimborne beyond the exquisite minster.   (pp. 77, 87, 96, 123.)

Wimborne St Giles (542), the seat of Lord Shaftesbury, has a very ornate church, with some interesting monuments. (p. 108.)

Winterborne Came (see **Came**).   The Winterborne parishes are named from a brook that only flows during the winter.   (p. 123.)

Woodsford Castle (see p. 104).

Wool (509), five miles west of Wareham, and north of Lulworth, for which it is the station.   A farmhouse here was formerly the manor-house of the Turbervilles.   (p. 109.)

Wyke Regis (1910) is now a suburb of Weymouth, of which it is the mother church.   There was a Celtic settlement here.   This part of the coast has been noted for its wrecks, and in 1805 was the scene of the wreck of the East Indiaman *Abergavenny*, whose captain was a brother of the poet Wordsworth. There is a fine Perp. church, rebuilt in 1455; also large torpedo-works.   (pp. 69, 98.)

Yetminster (551), four and a half miles south-east of Yeovil, has a good church, and is a pleasant unspoiled village.

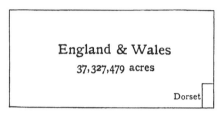

Fig. 1. The Area of Dorset, 624,031 acres, excluding
Water, compared with that of England and Wales

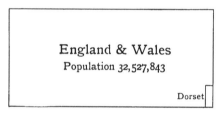

Fig. 2. The Population of Dorset (202,063) compared
with that of England and Wales in 1901

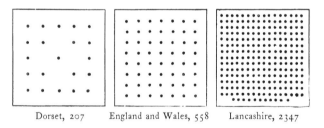

Dorset, 207   England and Wales, 558   Lancashire, 2347

Fig. 3. Comparative Density of Population
per square mile in 1901

(Each dot represents ten persons)

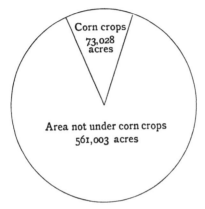

Fig. 4.   Proportionate Area under Corn Crops in
Dorset in 1908

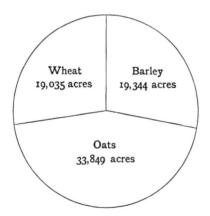

Fig. 5.   Proportionate Area of chief Cereals in
Dorset in 1908

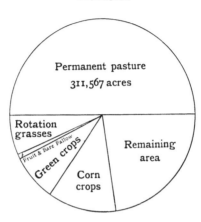

Fig. 6. Proportion of Permanent Pasture to other
Areas in Dorset in 1908

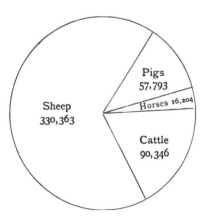

Fig. 7. Proportionate numbers of Live Stock
in Dorset in 1908

www.ingramcontent.com/pod-product-compliance
Ingram Content Group UK Ltd.
Pitfield, Milton Keynes, MK11 3LW, UK
UKHW042145280225
455719UK00001B/106